MW00587944

COORDINATE GEOMETRY

COORDINATE GEOMETRY

Luther Pfahler Eisenhart

DOVER PUBLICATIONS, INC.
Mineola, New York

Copyright

Bibliographical Note

This Dover edition, first published in 1960 and republished in 2005, is an unabridged republication of the first edition of the work, originally published by Ginn and Company, Boston, in 1939.

Library of Congress Cataloging-in-Publication Data

Eisenhart, Luther Pfahler, b. 1876.
Coordinate geometry / Luther Pfahler Eisenhart.
p. cm.
Originally published: Boston : Ginn and Co., c1939.
Includes index.
ISBN 0-486-44261-6 (pbk.)
1. Geometry, Analytic. 2. Coordinates. I. Title.

QA551.E58 2005
516.3—dc22

2004061384

Manufactured in the United States of America
Dover Publications, Inc., 31 East 2nd Street, Mineola, N.Y. 11501

Preface

"

The purposes and general plan of this book are set forth in the Introduction. Its practicability as a text has been tested and proved through use during two years in freshman courses in Princeton University. Each year it has been revised as a result of suggestions not only by the members of the staff but also by the students, who have shown keen interest and helpfulness in the development of the project.

An unusual feature of the book is the presentation of coordinate geometry in the plane in such manner as to lead readily to the study of lines and planes in space as a generalization of the geometry of the plane; this is done in Chapter 2. It has been our experience that the students have little, if any, difficulty in handling the geometry of space thus early in the course, in the way in which it is developed in this chapter.

We have also found that students who have studied determinants in an advanced course in algebra find for the first time in the definition and frequent use of determinants in this book an appreciation of the value and significance of this subject.

In the preparation and revisions of the text the author has received valuable assistance from the members of the staff at Princeton — in particular, Professors Knebelman and Tucker and Messrs. Tompkins, Daly, Fox, Titt, Traber, Battin, and Johnson; the last two have been notably helpful in the preparation of the final form of the text and in the reading of the proof. The Appendix to Chapter 1, which presents the relation between the algebraic foundations of coordinate geometry and axioms of Hilbert for Euclidean plane geometry, is due in the main to Professors Bochner and Church and Dr. Tompkins, particularly to the latter, at whose suggestion it was prepared and incorporated. The figures were drawn by Mr. J. H. Lewis.

It remains for me to express my appreciation of the courtesy and cooperation of Ginn and Company in the publication of this book.

LUTHER PFAHLER EISENHART
Princeton University

Contents

Contents

CHAPTER 3

Transformations of Coordinates

CHAPTER 4

The Conics · Locus Problems

CHAPTER 5

The Quadric Surfaces

Contents

Introduction

Coordinate geometry is so called because it uses in the treatment of geometric problems a system of coordinates, which associates with each point of a geometric figure a set of numbers — coordinates — so that the conditions which each point must satisfy are expressible by means of equations or inequalities ordinarily involving algebraic quantities and at times trigonometric functions. By this means a geometric problem is reduced to an algebraic problem, which most people can handle with greater ease and confidence. After the algebraic solution has been obtained, however, there remains its geometric interpretation to be determined; for, the problem is geometric, and algebra is a means to its solution, not the end. This method was introduced by René Descartes in *La Géométrie*, published in 1636; accordingly coordinate geometry is sometimes called Cartesian geometry. Before the time of Descartes geometric reasoning only was used in the study of geometry. The advance in the development of geometric ideas since the time of Descartes is largely due to the introduction of his method.

Geometry deals with spatial concepts. The problems of physics, astronomy, engineering, etc. involve not only space but usually time also. The method of attack upon these problems is similar to that used in coordinate geometry. However, geometric problems, because of the absence of the time element, are ordinarily simpler, and consequently it is advisable that the student first become familiar with the methods of coordinate geometry.

The aim of this book is to encourage the reader to *think mathematically*. The subject matter is presented as a unified whole, not as a composite of units which seem to have no relation to one another. Each situation is completely analyzed, and the various possibilities are all carried to their conclusion, for frequently the exceptional case (which often is not presented to a student) is the one that clarifies the general case — the idea epitomized in the old adage about an exception and a rule. Experience in analyzing a question fully and being careful in the handling of all possibilities is one of the great advantages of a proper study of mathematics.

Examples included in the text are there for the purpose of illustration and clarification of the text; there is no attempt to formulate a set of patterns for the reader so that the solution of exercises shall be a matter of memory alone without requiring him to think mathematically. However, it is not intended that he should develop no facility in mathematical techniques; rather these very techniques will have added interest when he understands the ideas underlying them. The reader may at first have some difficulty in studying the text, but if he endeavors to master the material, he will be repaid by finding coordinate geometry a very interesting subject and will discover that mathematics is much more than the routine manipulation of processes.

Since, as has been stated, coordinate geometry involves the use of algebraic processes in the study of geometric problems and also the geometric interpretation of algebraic equations, it is important that the reader be able not only to use algebraic processes but also to understand them fully, if he is to apply them with confidence. Accordingly in Chapter 1 an analysis is made of the solution of one and of two equations of the first degree in two unknowns; and in Chapter 2 are considered one, two, and three equations of the first degree in three unknowns. In order that the discussion be general and all-inclusive, literal coefficients are used in these equations. It may be that most, if not all, of the reader's experience in these matters has been with equations having numerical coefficients, and he may at first have some difficulty in dealing with literal coefficients. At times he may find it helpful in understanding the discussion to write particular equations with numerical coefficients, that is, to give the literal coefficients particular numerical values, thus supplementing illustrations of such equations which appear in the text. However, in the course of time he will find it unnecessary to do this, and in fact will prefer literal coefficients because of their generality and significance.

In the study of equations of the first degree in two or more unknowns determinants are defined and used, first those of the second order and then those of the third and higher orders, as they are needed. Ordinarily determinants are defined and studied first in a course in algebra, but it is a question whether one ever appreciates their value and power until one sees them used effectively in relation to geometric problems.

The geometry of the plane is presented in Chapter 1 in such form that the results may be generalized readily to ordinary space and to spaces of four and more dimensions, as is done in Chapter 2.

Some of the exercises are a direct application of the text so that the reader may test his understanding of a certain subject by applying it to a particular problem and thus also acquire facility in the appropriate techniques; others are of a theoretical character, in the solution of which the reader is asked to apply the principles of the text to the establishment of further theorems. Some of these theorems extend the scope of the text, whereas others complete the treatment of the material in the text.

Of general interest in this field are A. N. Whitehead's *Introduction to Mathematics*, particularly Chapter 8, and E. T. Bell's *Men of Mathematics*, Chapter 3.

"

COORDINATE GEOMETRY

CHAPTER 1

Points and Lines in the Plane

"

II

1. The Equation of the First Degree in x and y

In his study of algebra the reader has no doubt had experience in finding solutions of equations of the first degree in two unknowns, x and y, as for example $x - 2y + 3 = 0$. Since we shall be concerned with the geometric interpretation of such equations, a thorough understanding of them is essential. We therefore turn our attention first of all to a purely algebraic study of a single equation in x and y, our interest being to find out what statements can be made about a general equation of the first degree, which thus will apply to any such equation without regard to the particular coefficients it may have. Accordingly we consider the equation

$$ax + by + c = 0, \tag{1.1}$$

where a, b, and c stand for arbitrary numbers, but definite in the case of a particular equation. We say that a value of x and a value of y constitute a *solution* of this equation if the left-hand side of this equation reduces to zero when these values are substituted. Does such an equation have a solution whatever be the coefficients? The question is not as trivial as may appear at first glance, and the method of answering it will serve as an example of the type of argument used repeatedly in later sections of this book.

We consider first the case when a in (1.1) is not equal to zero, which we express by $a \neq 0$. If we substitute any value whatever for y in (1.1) and transpose the last two terms to the right-hand side of the equation, which involves changing their signs, we may divide through by a and obtain the value $x = -\dfrac{by + c}{a}$. This value of x and the chosen value of y satisfy the equation, as one sees by substitution; hence any value of y and the resulting value of x constitute a solution. Since y may be given any value and then x is determined, we say that when $a \neq 0$ there is an *endless number of solutions* of the equation.

The above method does not apply when $a = 0$, that is, to the equation

$$0x + by + c = 0,$$

since division by zero is not an allowable process. The reader may have been told that it is allowable, and that any number, for example 2, divided by zero is infinity; but infinity defined in this manner is a concept quite different from ordinary numbers, and an understanding of the concept necessitates an appropriate knowledge of the theory of limits. If now $b \neq 0$, the above equation may be solved for y; that is, $y = -c/b$. For this equation also there is an endless number of solutions, for all of which y has the value $-c/b$ and x takes on arbitrary values. Usually the above equation is written $by + c = 0$, which does not mean that x is equal to zero (a mistake frequently made), but that the coefficient of x is zero.

If $b \neq 0$, no matter what a is, equation (1.1) may be solved for y with the result $y = -\dfrac{ax + c}{b}$, the value of y corresponding to any choice of x being given by this expression. Hence we have the theorem

[1.1] *An equation of the first degree in two unknowns in which the coefficient of at least one of the unknowns is not equal to zero has an endless number of solutions.*

There remains for consideration the case when $a = 0$ and $b = 0$; that is, the equation

$$0x + 0y + c = 0.$$

Evidently there are no solutions when $c \neq 0$, and when $c = 0$ any value of x and any value of y constitute a solution. The reader may say that in either case this is really not an equation of the first degree in x and y, and so why consider it. It is true that one would not start out with such an equation in formulating a set of one or more equations of the first degree to express in algebraic form the conditions of a geometric problem, but it may happen that, having started with several equations, and carrying out perfectly legitimate processes, one is brought to an equation of the above type; that is, equations of this type do arise and consequently must be considered. In fact, this situation arises in § 9, and the reader will see there how it is interpreted. However, when in this chapter we are deriving theorems concerning equations of the first degree in x and y, we exclude the *degenerate* case when the coefficients of both x and y are zero.

Any solution of equation (1.1) is also a solution of the equation $k(ax + by + c) = 0$, where k is any constant different from zero. Moreover, any solution of this equation is a solution of (1.1); for, if we are seeking the conditions under which the product of two quantities shall be equal to zero and one of the quantities is different from zero by hypothesis, then we must seek under what conditions the other quantity is equal to zero. Accordingly we say that two equations differing only by a constant factor are *not essentially different*, or are *not independent*, or that the two equations are *equivalent*. In view of this discussion it follows that a common factor, if any, of all the coefficients of an equation can be divided out, or the signs of all the terms of an equation can be changed, without affecting the solutions of the equation — processes which the reader has used even though he may not have thought how to justify their use.

It should be remarked that the values of the coefficients in equation (1.1) are the important thing, because they fix y when x is chosen and vice versa. This is seen more clearly when we take a set of values x_1, y_1 and seek the equation of which it is a solution; this is the inverse of the problem of finding solutions of a given equation. Here the subscript 1 of x_1 and y_1 has nothing to do with the values of these quantities. It is a means of denoting a particular solution of an equation, whereas x and y without any subscript denote any solution whatever.

If x_1 and y_1 is to be a solution of (1.1), the coefficients a, b, and c must be such that

$$ax_1 + by_1 + c = 0. \tag{1.2}$$

On solving this equation for c and substituting in (1.1), we obtain the equation

$$ax + by - ax_1 - by_1 = 0, \tag{1.3}$$

which is of the form (1.1), where now c is $-(ax_1 + by_1)$, a constant for any values of a and b, x_1 and y_1 being given constants.

When equation (1.3) is rewritten in the form

$$a(x - x_1) + b(y - y_1) = 0, \tag{1.4}$$

it is seen that $x = x_1$ and $y = y_1$ is a solution of (1.3) whatever be a and b. Since a and b can take any values in (1.4), we see that an equation of the first degree in x and y is not completely determined (that is, a, b, and c are not fixed) when one solution is given.

Suppose then that we require that a different set of quantities, x_2 and y_2, be also a solution, the subscript 2 indicating that it is a second solution. On replacing x and y in (1.4) by x_2 and y_2, we obtain

$$a(x_2 - x_1) + b(y_2 - y_1) = 0. \tag{1.5}$$

Since we are dealing with two different solutions, either $x_2 \neq x_1$ or $y_2 \neq y_1$; we assume that $x_2 \neq x_1$ and solve (1.5) for a, with the result

$$a = -b\frac{y_2 - y_1}{x_2 - x_1}. \tag{1.6}$$

On substituting this value of a in (1.4), we obtain

$$b\left[-(x - x_1)\frac{y_2 - y_1}{x_2 - x_1} + (y - y_1)\right] = 0.$$

This equation is satisfied if $b = 0$, but then from (1.5) we have $a(x_2 - x_1) = 0$. Since $x_2 \neq x_1$, we must have $a = 0$, contrary to the hypothesis that a and b are not both equal to zero. Therefore since b cannot be zero, it may be divided out of the above equation, and what remains may be written

$$y - y_1 = \frac{y_2 - y_1}{x_2 - x_1}(x - x_1). \tag{1.7}$$

Multiplying both sides by $x_2 - x_1$, the resulting equation may be written

$$(y_2 - y_1)x - (x_2 - x_1)y + (x_2y_1 - x_1y_2) = 0, \tag{1.8}$$

which is of the form (1.1), since x_1, y_1, x_2, and y_2 are fixed numbers.

Thus far we have assumed that $x_2 \neq x_1$; if now $x_2 = x_1$ we cannot have also $y_2 = y_1$, since the two solutions are different, and consequently from (1.5) we have $b = 0$. In this case equation (1.4) becomes $a(x - x_1) = 0$. We observe that this equation differs only by the constant factor a from $x - x_1 = 0$. And the same is true of the form which (1.8) takes when we put $x_2 = x_1$, namely, $(y_2 - y_1)(x - x_1) = 0$. But we have remarked before that a common factor of all the coefficients does not affect the solutions of the equation.

Accordingly we have the theorem

[1.2] *Equation* (1.8) *is an equation of the first degree in* x *and* y *which has the two solutions* x_1, y_1 *and* x_2, y_2.

We say "an equation" and not "the equation" because any constant multiple of equation (1.8) also has these solutions; in this sense an equation is determined by two solutions to within an arbitrary constant factor.

As a result of the discussion leading up to theorem [**1.2**] we have the theorem

[1.3] *Although an equation of the first degree in* x *and* y *admits an endless number of solutions, the equation is determined to within an arbitrary constant factor by two solutions, that is, by two sets of values of* x *and* y; *for the solutions* x_1, y_1 *and* x_2, y_2 *the equation is equivalent to* (1.8).

NOTE. In the numbering of an equation, as (1.5), the number preceding the period is that of the section in which the equation appears, and the second number specifies the particular equation. The same applies to the number of a theorem, but in this case a bracket is used instead of a parenthesis.

EXERCISES

1. What values must be assigned to a, b, and c in equation (1.1) so that the resulting equation is $x = 0$; so that it is equivalent to this equation?

2. Find two solutions of the equation

$$2x - 3y + 6 = 0, \tag{i}$$

and show that equation (1.8) for these two solutions is equivalent to equation (i).

3. Show that the equation $x - 2y + 3 = 0$ does not have two different solutions for both of which $x = 2$; find an equation of the first degree in x and y for which this is true.

4. Show that it follows from (1.1) and (1.8) that

$$a : b : c = y_2 - y_1 : x_1 - x_2 : x_2y_1 - x_1y_2,$$

where x_1, y_1 and x_2, y_2 are solutions of (1.1).

5. Criticize the following statements:

a. In obtaining solutions of an equation $ax + by + c = 0$, x may take any value if $a = 0$, and only in this case.

b. $x - 3 = 0$, as an equation in x and y, has the solution $x = 3$, $y = 0$, and this is the only solution.

2. Cartesian Coordinates in the Plane

Having studied equations of the first degree in two unknowns, we turn now to the geometric interpretation of the results of this study. This is done by the introduction of coordinates, which serve as the bridge from algebra to geometry. It is a bridge with two-way traffic; for also by means of coordinates geometric problems may be given algebraic form. This use of coordinates was Descartes's great contribution to mathematics, which revolutionized the study of geometry.

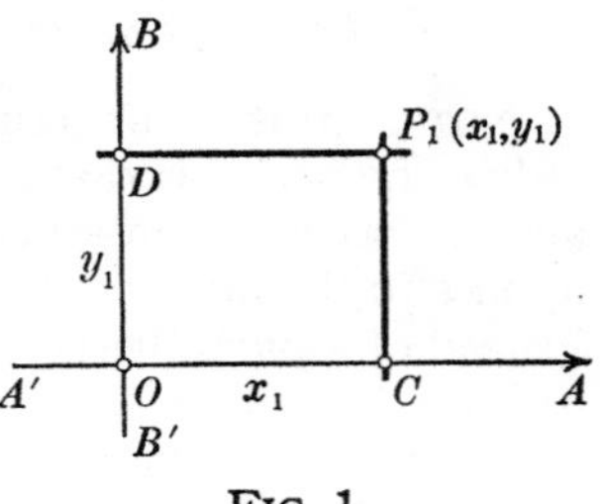

FIG. 1

As basis for the definition of coordinates, we take two lines perpendicular to one another, as $A'A$ and $B'B$ in Fig. 1, which are called the *x-axis* and *y-axis* respectively; their intersection O is called the *origin*. Suppose that x_1 and y_1 are a pair of numbers. If x_1 is positive, we lay off on OA from O a length OC equal to x_1 units and draw through C a line parallel to $B'B$; if x_1 is negative, we lay off from O a length equal to $-x_1$ units on OA', and draw through the point so determined a line parallel to $B'B$. Then, starting from O we lay off on OB a length OD equal to y_1 units if y_1 is positive, or on OB' a length $-y_1$ units if y_1 is negative, and through the point so determined draw a line parallel to $A'A$. These lines so drawn meet in a point P_1, which is called the *graph* of the pair x_1, y_1; we say that P_1 is the point (x_1, y_1), and x_1, y_1 are called the *coordinates* of P_1; also x_1 is called the *abscissa* of P_1 and y_1 the *ordinate*. Evidently x_1 is the distance of P_1 from the y-axis (to the right if x_1 is positive, to the left if x_1 is negative) and y_1 is the distance of P_1 from the x-axis.

In specifying a line segment OC, the first letter O indicates the point from which measurement begins, and the last letter C the point to which measurement is made. Accordingly we have $CO = -OC$, because for CO measurement is in the direction opposite to that for OC. (The question of sign may be annoying, but in many cases it is important; there are also cases when it is not important, and the reader is expected to discriminate between these cases.) If we take two points $C_1(x_1, 0)$ and $C_2(x_2, 0)$ on the x-axis, we see that the magnitude and sign of the segment C_1C_2 is $x_2 - x_1$, since $C_1C_2 = OC_2 - OC_1$. When C_1 and C_2 lie on the same side of O, $OC_2 - OC_1$ is the difference of two lengths; when they are on opposite sides of O, it is the sum of two lengths. (The reader does not have to worry about this, since $x_2 - x_1$ takes care of all of it.) C_1C_2 is called the *directed distance* from C_1 to C_2.

We observe that the coordinate axes divide the plane into four compartments, which are called *quadrants*. The quadrant formed by the positive x-axis and positive y-axis is called the *first* quadrant; the one to the left of the positive y-axis (and above the x-axis), the *second* quadrant; the ones below the negative x-axis and the positive x-axis, the *third* and *fourth* quadrants respectively.

By definition the *projection of a point upon a line* is the foot of the perpendicular to the line from the point. Thus C and D in Fig. 1 are the projections of the point P_1 upon the x-axis and y-axis respectively. The *projection of the line segment* whose end points are P_1 and P_2 *upon a line* is the line segment whose end points are the projections of P_1 and P_2 on the line. Thus in Fig. 1 the line segment OC is the projection of the line segment DP_1 upon the x-axis.

Two points P_1 and P_2 are said to be *symmetric with respect to a point* when the latter bisects the line segment P_1P_2; *symmetric with respect to a line* when the latter is perpendicular to the line segment P_1P_2 and bisects it. Thus the points (3, 2) and $(-3, -2)$ are symmetric with respect to the origin, and $(3, 2)$ and $(3, -2)$ are symmetric with respect to the x-axis.

The reader should acquire the habit of drawing a reasonably accurate graph to illustrate a problem under consideration. A carefully made graph not only serves to clarify the geometric interpretation of a problem but also may serve as a valuable check on the accuracy of the algebraic work. Engineers, in particular, often use graphical methods, and for many purposes numerical results obtained graphically are sufficiently accurate. However, the reader should never forget that graphical results are at best only approximations, and of value only in proportion to the accuracy with which the graphs are drawn. Also in basing an argument upon a graph one must be sure that the graph is really a picture of the conditions of the problem (see Chapter 3 of W. W. R. Ball's *Mathematical Essays and Recreations*, The Macmillan Company).

EXERCISES

1. How far is the point $(4, -3)$ from the origin? What are the coordinates of another point at the same distance from the origin? How many points are there at this distance from the origin, and how would one find the coordinates of a given number of such points?

2. What are the coordinates of the point halfway between the origin and $(-4, 6)$? What are the coordinates of the point such that $(-4, 6)$ is halfway between it and the origin?

3. Where are the points in the plane for which $x > y$ (that is, for which x is greater than y)? Where are the points for which $x^2 + y^2 < 4$?

4. Where are the points for which $0 < x \leq 1$ and $0 < y \leq 1$ (the symbol $\leq$ meaning "less than or equal to")? Where are the points for which $0 < y < x < 1$?

5. Where are the points for which $xy = 0$?

6. What are the lengths of the projections upon the x-axis and y-axis of the line segment whose end points are $(1, 2)$ and $(-3, 4)$?

7. A line segment with the origin as an end point is of length l. What are its projections upon the x-axis and y-axis in terms of l and the angles the segment makes with the axes?

8. Given the four points $P_1(1, -2)$, $P_2(3, 4)$, $P_3(-5, 1)$, and $P_4(0, 3)$, show that the sum of the projections of the line segments

P_1P_2, P_2P_3, and P_3P_4 on either the x-axis or the y-axis is equal to the projection of P_1P_4 on this axis. Is this result true for any four points; for any number of points?

9. Given the point $(2, -3)$, find the three points which are symmetric to it with respect to the origin, and to the x-axis and y-axis respectively.

3. Distance between Two Points. Direction Numbers and Direction Cosines. Angle between Directed Line Segments

Consider two points $P_1(x_1, y_1)$ and $P_2(x_2, y_2)$, and the line segment P_1P_2 joining them, as in Fig. 2, in which the angle at Q is a right angle. The square of the distance between the points, that is, the square of the line segment P_1P_2, is given by

$$(P_1P_2)^2 = (P_1Q)^2 + (QP_2)^2 = (x_2 - x_1)^2 + (y_2 - y_1)^2.$$

Hence we have the theorem

[3.1] *The distance between the points* (x_1, y_1) *and* (x_2, y_2) *is*

$$\sqrt{(x_2-x_1)^2+(y_2-y_1)^2}. \tag{3.1}$$

The length and sign of the directed segments P_1Q and P_1R in Fig. 2 are given by

$$P_1Q = x_2 - x_1, \quad P_1R = y_2 - y_1, \tag{3.2}$$

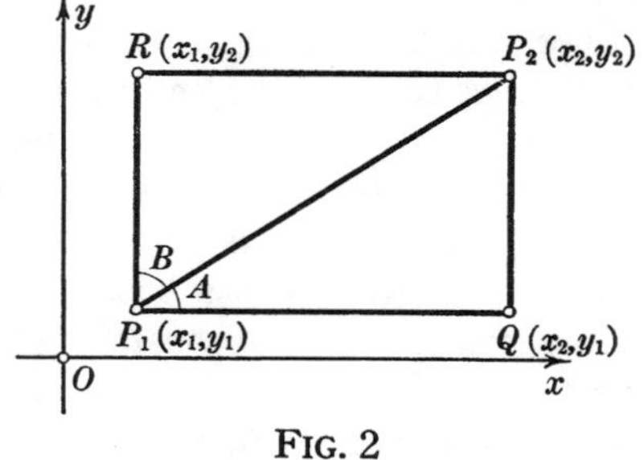

FIG. 2

no matter in which quadrant P_1 lies and in which P_2, as the reader will see when he draws other figures. These numbers determine the rectangle of which P_1P_2 is a diagonal, and consequently determine the direction of P_1P_2 relative to the coordinate axes. They are called *direction numbers* of the line segment P_1P_2. In like manner $x_1 - x_2$ and $y_1 - y_2$ are direction numbers of the line segment P_2P_1. Thus a line segment has two sets of direction numbers, each associated with a *sense along the segment* and either determining the direction of the segment relative to the coordinate axes. But a *sensed* line segment, that is, a segment with an assigned sense, has a single set of direction numbers.

Any other line segment parallel to P_1P_2 and having the same length and sense as P_1P_2 has the same direction numbers as P_1P_2; for, this new segment determines a rectangle equal in every respect to the one for P_1P_2. This means that the differences of the x's and the y's of the end points are equal to the corresponding differences for P_1 and P_2. Since one and only one line segment having given direction numbers can be drawn from a given point, we have that a sensed line segment is completely determined by specifying its *initial* point and its direction numbers.

There is another set of numbers determining the direction of a line segment, called the *direction cosines,* whose definition involves a convention as to the positive sense along the segment. If the segment is parallel to the x-axis, we say that its *positive sense* is that of the positive direction of the x-axis. If the segment is not parallel to the x-axis, we make the convention that upward along the segment is the *positive sense* on the segment. This is in agreement with the sense already established on the y-axis. In Fig. 2 the distance P_1P_2 is a positive number, being measured in the positive sense, and the distance P_2P_1 is a negative number (the numerical, or *absolute,* value of these numbers being the same), just as distances measured on the x-axis to the right, or left, of a point on the axis are positive, or negative. When the positive sense of a line segment is determined by this convention, we refer to it as a *directed* line segment.

By definition the *direction cosines of a line segment are the cosines of the angles which the positive direction of the line segment makes with the positive directions of the x-axis and y-axis respectively, or, what is the same thing, with the positive directions of lines parallel to them.* They are denoted respectively by the Greek letters λ (lambda) and μ (mu). Thus, in Fig. 2, for the line segment P_1P_2

$$\lambda = \cos A, \qquad \mu = \cos B, \tag{3.3}$$

and also

$$\begin{aligned} P_1Q &= P_1P_2 \cos A = P_1P_2\lambda, \\ P_1R &= P_1P_2 \cos B = P_1P_2\mu. \end{aligned} \tag{3.4}$$

If we denote the positive distance P_1P_2 by d, we have from (3.4) and (3.2)

$$x_2 - x_1 = d\lambda, \qquad y_2 - y_1 = d\mu. \tag{3.5}$$

If we imagine P_1 and P_2 interchanged in Fig. 2, this does not alter λ and μ, since their values depend only upon the direction of the segment relative to the coordinate axes; and consequently equations (3.5) become

$$x_1 - x_2 = \bar{d}\lambda, \qquad y_1 - y_2 = \bar{d}\mu,$$

where $\bar{d}$ is the distance P_2P_1 in the new figure and is positive. Hence equations (3.5) hold also if d is negative, that is, when P_1 is above P_2 on the line and the distance P_1P_2 is negative.

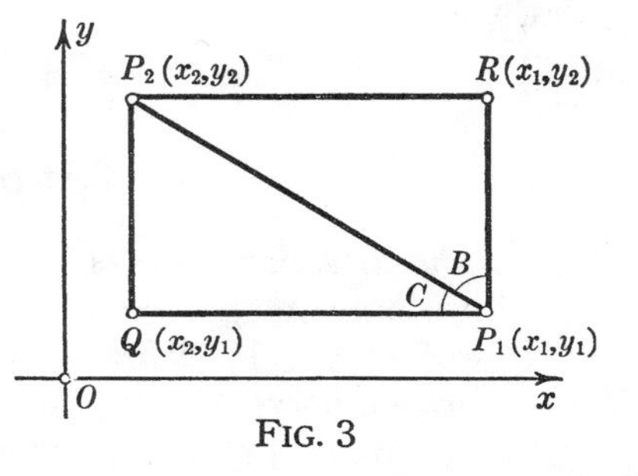

FIG. 3

We consider now the difference, if any, in the above results when the positive sense of the segment makes an obtuse angle with the positive sense of the x-axis. Such a situation is shown in Fig. 3, where this angle is $180° - C$. In this case we have

$$P_1Q = -\,QP_1 = -\,P_1P_2 \cos C = P_1P_2\lambda,$$
$$P_1R = P_1P_2 \cos B = P_1P_2\mu,$$

which are the same as (3.4); and consequently equations (3.5) hold in every case, with the understanding that d is the directed distance P_1P_2. Thus a line segment is completely determined by an end point, its direction cosines, and its length and sense relative to the given end point.

Two line segments are parallel if their direction cosines are equal, in which case their direction numbers are proportional, as follows from (3.4). As a means of distinguishing line segments with the same direction but with opposite senses, some writers reserve the term *parallel* for segments with the same direction and the same sense, and use the term *antiparallel* for the case when the senses are opposite; such distinction is necessary in mechanics, for example.

From the definition of the positive direction along a segment it follows that when P_1P_2 is not parallel to the x-axis μ is positive, whereas λ may take any value between -1 and $+1$. When the segment is parallel to the x-axis, $\mu = 0$ and $\lambda = 1$; in fact, the direction cosines of the x-axis and y-axis are 1, 0 and 0, 1 respectively.

From the definition of d and Theorem [3.1], we have

$$(x_2 - x_1)^2 + (y_2 - y_1)^2 = d^2.$$

When the expressions from (3.5) are substituted in this equation, we obtain

$$\lambda^2 + \mu^2 = 1. \tag{3.6}$$

Accordingly we have the first part of the theorem

[3.2] *The direction cosines* λ, μ *of any line segment satisfy the equation* $\lambda^2 + \mu^2 = 1$; μ *is never negative; when* $\mu > 0$, $-1 < \lambda < 1$; *when* $\mu = 0$, $\lambda = 1$; *and, conversely, any two numbers* λ *and* μ *satisfying these conditions are direction cosines of a line segment.*

To prove the second part of the theorem we take any point $P_1(x_1, y_1)$ and an arbitrary number d, and determine numbers x_2 and y_2 from equations (3.5). Then λ and μ are direction cosines of the line segment P_1P_2, where P_2 is the point (x_2, y_2) and d is the directed distance P_1P_2; and the theorem is proved.

Consider now in connection with the line segment P_1P_2 another line segment P_1P_3, where P_3 is the point (x_3, y_3). We denote by θ (theta) the angle formed at P_1 by these sensed segments; and we note that if they have the same direction, $\theta = 0°$ or $180°$ according as the segments have the same or opposite sense. We consider the case when θ is not equal to $0°$ or $180°$, and denote by λ_1, μ_1 and λ_2, μ_2 the direction cosines of the segments P_1P_2 and P_1P_3 respectively, and by d_1 and d_2 the directed distances P_1P_2 and P_1P_3. We draw, as in Fig. 4, from the origin two line segments, OP_1' and OP_2', parallel to and of the same directed lengths as P_1P_2 and P_1P_3 respectively. The direction cosines of these line segments are λ_1, μ_1 and

λ_2, μ_2, and the coordinates of P_1' and P_2' are $d_1\lambda_1$, $d_1\mu_1$ and $d_2\lambda_2$, $d_2\mu_2$ respectively, as follows from equations analogous to (3.5) with $x_1 = y_1 = 0$; and the angle of these segments is equal to the angle θ of the original line segments. By the Law of Cosines of plane trigonometry applied to the triangle in Fig. 4, in which l, l_1, and l_2 denote the lengths of the sides of the triangle, as distinguished from directed distances, we have

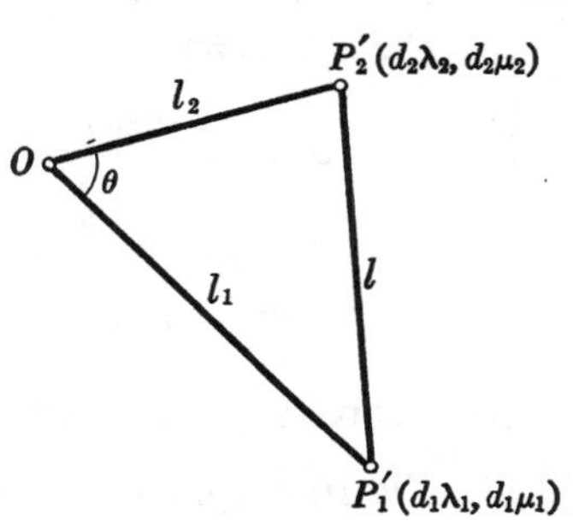

FIG. 4

(3.7) $$l^2 = l_1^2 + l_2^2 - 2\, l_1 l_2 \cos\theta.$$

By means of Theorems [3.1] and [3.2] we have

(3.8) $$\begin{aligned} l^2 &= (d_2\lambda_2 - d_1\lambda_1)^2 + (d_2\mu_2 - d_1\mu_1)^2 \\ &= d_2^2(\lambda_2^2 + \mu_2^2) + d_1^2(\lambda_1^2 + \mu_1^2) - 2\, d_1 d_2(\lambda_1\lambda_2 + \mu_1\mu_2) \\ &= d_2^2 + d_1^2 - 2\, d_1 d_2(\lambda_1\lambda_2 + \mu_1\mu_2). \end{aligned}$$

From the definition of l_1 and l_2 it follows that l_1 is equal to d_1 or $-d_1$ according as d_1 is positive or negative, and similarly for l_2; from this it follows that $l_1^2 = d_1^2$ and $l_2^2 = d_2^2$. When we equate the expressions (3.7) and (3.8) for l^2, and simplify the resulting equation, we obtain

$$l_1 l_2 \cos\theta = d_1 d_2(\lambda_1\lambda_2 + \mu_1\mu_2).$$

If d_1 and d_2 are both positive or both negative, $l_1 l_2 = d_1 d_2$; and if d_1 and d_2 differ in sign, $l_1 l_2 = -d_1 d_2$. Hence we have the theorem

[3.3] *The angle θ between two line segments whose direction cosines are λ_1, μ_1 and λ_2, μ_2 is given by*

(3.9) $$\cos\theta = e(\lambda_1\lambda_2 + \mu_1\mu_2),$$

where e is $+1$ or -1 according as the two segments have the same sense (both positive or both negative) or opposite senses.

Here, and throughout the book, we use e in place of the sign $\pm$, because it enables one to state more clearly when the sign is $+$ and when $-$.

From Theorem [3.3] and equations (3.5) we have

[3.4] *The angle θ between the directed line segments P_1P_2 and P_1P_3 is given by*

$$\cos\theta = \frac{(x_2 - x_1)(x_3 - x_1) + (y_2 - y_1)(y_3 - y_1)}{l_1 l_2}, \tag{3.10}$$

where l_1 and l_2 are the lengths (not directed distances) of the segments P_1P_2 and P_1P_3 respectively.

As a corollary we have

[3.5] *The line segments from the point (x_1, y_1) to the points (x_2, y_2) and (x_3, y_3) are perpendicular, if and only if*

$$(x_2 - x_1)(x_3 - x_1) + (y_2 - y_1)(y_3 - y_1) = 0. \tag{3.11}$$

The phrase "if and only if" used in this theorem (and throughout the book) is a way of stating that both a theorem with "if" alone and its converse are true; that is, a statement that A is true if and only if B holds means that A is true if B holds, and B is true if A holds.

EXERCISES

1. Show that the triangle with the vertices $(6, -5)$, $(2, -4)$, and $(5, -1)$ is isosceles.

2. Find the point on the x-axis which is equidistant from $(0, -1)$ and $(3, 3)$.

3. Find x_1 and y_1 such that $P_1(x_1, y_1)$, $(0, 0)$, and $(3, -4)$ are the vertices of an equilateral triangle.

4. Find the direction numbers and direction cosines of the line segment joining $P_1(1, -2)$ to $P_2(4, 2)$.

5. Find the coordinates of the point $P(x, y)$ so that the line segment P_1P has the same direction and is twice as long as the line segment in Ex. 4.

6. For what value of a is the line segment from the point $P_1(2, -1)$ to the point $P(3, a)$ perpendicular to the line segment P_1P_2, where P_2 is the point $(5, 1)$?

7. Find the lengths and direction cosines of the sides of the triangle whose vertices are $(3, -6)$, $(8, -2)$, and $(-1, -1)$; also the cosines of the angles of the triangle.

8. Prove that $(2, 1)$, $(0, 0)$, $(-1, 2)$, and $(1, 3)$ are the vertices of a rectangle.

9. Find the condition to be satisfied by the coordinates of the points $P_1(x_1, y_1)$ and $P_2(x_2, y_2)$ in order that the line segment P_1P_2 shall subtend a right angle at the origin.

10. Derive equation (3.10) directly by applying the Law of Cosines to the triangle with vertices at the points (x_1, y_1), (x_2, y_2), (x_3, y_3).

11. Find the coordinates of a point $P(x, y)$ so that the segment P_1P has the same direction as the segment P_1P_2 with the end points (x_1, y_1) and (x_2, y_2). How many points P possess this property, and where are they?

12. Find the coordinates of a point $P(x, y)$ so that the segment P_1P is perpendicular to the segment P_1P_2 with the end points (x_1, y_1) and (x_2, y_2). How many points P possess this property, and where are they?

4. Internal and External Division of a Line Segment

In this section we determine the coordinates of the point $P(x, y)$ which divides the line segment P_1P_2 in the ratio h_1/h_2, that is, such that

$$\frac{P_1P}{PP_2} = \frac{h_1}{h_2}. \tag{4.1}$$

Since the triangles P_1QP and PRP_2 in Fig. 5 are similar, we have, by a theorem of plane geometry, that corresponding sides of these triangles are proportional; thus we have

$$\frac{P_1P}{PP_2} = \frac{P_1Q}{PR} = \frac{QP}{RP_2}. \tag{4.2}$$

Since
$$P_1Q = x - x_1,$$
$$PR = x_2 - x,$$
$$QP = y - y_1,$$
$$RP_2 = y_2 - y,$$

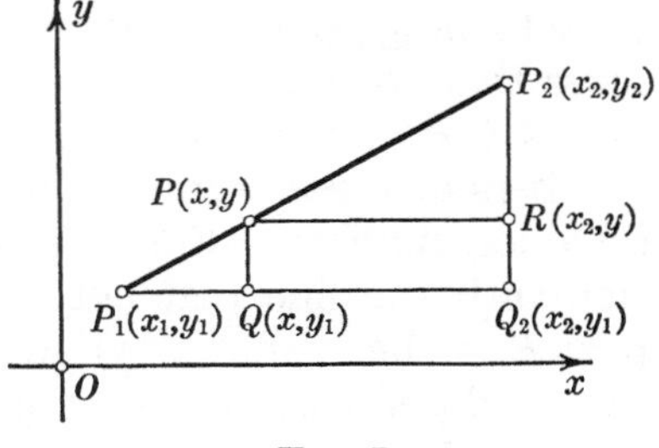

FIG. 5

it follows from (4.1) and (4.2) that x and y are such that

$$\frac{h_1}{h_2} = \frac{x - x_1}{x_2 - x} = \frac{y - y_1}{y_2 - y}. \tag{4.3}$$

From the equation of the first two terms we have

$$\frac{h_1}{h_2}(x_2 - x) = x - x_1, \tag{4.4}$$

from which, on solving for x, we obtain

$$x = \frac{x_1 + \dfrac{h_1}{h_2} x_2}{1 + \dfrac{h_1}{h_2}} = \frac{h_2 x_1 + h_1 x_2}{h_2 + h_1}. \tag{4.5}$$

Proceeding in like manner with the equation consisting of the first and third terms in (4.3), we obtain

$$y = \frac{h_2 y_1 + h_1 y_2}{h_2 + h_1}. \tag{4.6}$$

If the points P_1, P_2, and P are on a line parallel to the x-axis or are on the x-axis, the situation is expressed by (4.4), from which (4.5) follows; similarly if these points are on a line parallel to the y-axis or are on the y-axis, we have (4.6).

In order to obtain the coordinates of the mid-point, we put $h_1 = h_2 = 1$ in (4.5) and (4.6), and have the theorem

[4.1] *The coordinates of the mid-point of the line segment joining the points* (x_1, y_1) *and* (x_2, y_2) *are*

$$\tfrac{1}{2}(x_1 + x_2), \quad \tfrac{1}{2}(y_1 + y_2). \tag{4.7}$$

Returning to the consideration of equations (4.3), we observe that they express the condition that the ratio of the segments P_1P and PP_2 is equal to the ratio of the first direction numbers of these segments, and also to the ratio of the second direction numbers. Suppose then that we consider in connection with the line segment P_1P_2 the line segment P_2P, where now P is the point (x, y) such that P_2P has the same direction and sense as P_1P_2; then the ratio P_1P/PP_2 is a negative number, since the segments P_1P and PP_2 have opposite sense, but their algebraic sum is equal to P_1P_2. In this case we again have equations (4.3), in which, however, the quantity h_1/h_2 is a negative number, that is, either h_1 or h_2 may be taken as negative and the other positive; but the numerical value of h_1 is greater than the numerical value of h_2, since the length of P_1P is greater than the length of PP_2. With this understanding about h_1 and h_2, the coordinates of P in terms of those of P_1 and P_2 are given by (4.5) and (4.6).

It is customary to denote the numerical, or *absolute*, value of a number a by $|a|$; thus $|2| = 2$, $|-2| = 2$. Then the above statement about h_1 and h_2 is expressed by $|h_1| > |h_2|$. On the other hand, if we have equations (4.5) and (4.6) with h_1 and h_2 differing in sign, and $|h_1| < |h_2|$, this means that x and y are the coordinates of a point P lying below P_1 on the line through P_1 and P_2, as the reader will see by considering the above definition of P for the case $|h_1| < |h_2|$, and drawing a figure. In accordance with custom we say that in these cases $P(x, y)$ divides the line segment P_1P_2 *externally* in the ratio h_1/h_2; and that when P is a point of the segment, it divides the segment P_1P_2 *internally*. Accordingly we have the theorem

[4.2] *The equations*

$$x = \frac{h_2x_1 + h_1x_2}{h_2 + h_1}, \qquad y = \frac{h_2y_1 + h_1y_2}{h_2 + h_1} \tag{4.8}$$

give the coordinates of the point P dividing the line segment P_1P_2, with end points $P_1(x_1, y_1)$, $P_2(x_2, y_2)$, in the ratio h_1/h_2, internally when h_1 and h_2 have the same sign and externally when h_1 and h_2 have opposite signs; in either case P lies nearer P_1 or P_2 according as $|h_1|$ is less or greater than $|h_2|$.

EXERCISES

1. Find the coordinates of the points dividing the line segment between the points (7, 4) and (5, − 6) in the ratios 2/3 and − 2/3.

2. Find the coordinates of the two points which trisect the line segment between the points (3, − 2) and (− 3, 1).

3. In what ratio does the point (3, − 2) divide the line segment between the points (5, 2) and (6, 4)?

4. Find the coordinates of the point $P(x, y)$ on the line through the points $P_1(5, -3)$ and $P_2(2, 1)$ such that P_2 bisects the segment P_1P.

5. Show that the medians of the triangle with vertices $(-1, 2)$, $(3, 4)$, and $(5, 6)$ meet in a point by finding for each median the coordinates of the point twice as far from the corresponding vertex as from the opposite side, distances being measured along the median.

6. Is there a point on the line through $P_1(x_1, y_1)$ and $P_2(x_2, y_2)$ dividing the line segment P_1P_2 in the ratio $h_1/h_2 = -1$?

7. Where are the points in the plane for which $|x| > 2$ and $|y| < 3$?

8. Show that the points (x_1, y_1) and (x_2, y_2) are collinear with the origin if and only if their coordinates are proportional.

9. Show that the points (x_1, y_1), (x_2, y_2), and (x_3, y_3) are collinear when there are three numbers k_1, k_2, k_3, all different from zero, such that $k_1 + k_2 + k_3 = 0$, $k_1x_1 + k_2x_2 + k_3x_3 = 0$, $k_1y_1 + k_2y_2 + k_3y_3 = 0$.

10. Interpret the following statements:

a. For h_1 and h_2 both positive x defined by (4.8) is the weighted average of x_1 and x_2 with the respective weights h_2 and h_1, and according as h_2 is greater than h_1 or less than h_1 the weighted average is nearer x_1 than x_2 or vice versa.

b. x and y given by (4.8) are the coordinates of the center of gravity of masses h_2 at P_1 and h_1 at P_2.

11. What are the coordinates of the center of gravity of masses m_1, m_2, and m_3 at the points (x_1, y_1), (x_2, y_2), and (x_3, y_3); of n masses at n different points? (Use mathematical induction.)

5. An Equation of a Line. Parametric Equations of a Line

We have seen how any two numbers x and y define a point in the plane, called the *graph* of the *ordered pair* of numbers. Reversing the process, we have that any point in the plane is the graph of an ordered pair of numbers, the coordinates of the point, obtained by drawing through the point lines parallel to the coordinate axes and taking for these numbers the magnitudes of the intercepted segments with appropriate signs, as OC and OD in Fig. 1. An equation in x and y imposes a restriction on x and y, so that only one of these may take arbitrary values and then the other is determined by the equation. *The graph of an equation is the locus (place) of all points whose coordinates are solutions of the equation, in other words, the locus of the graphs of ordered pairs of numbers satisfying the equation.* Each of the points of this locus possesses a geometric property common to all the points of the locus, and no other points; this property is expressed algebraically by the equation. Just what this geometric property is in the case of a particular equation is one of the interesting questions of coordinate geometry. The converse problem is that of finding an equation whose solutions give the coordinates of every point of a locus when the locus is defined geometrically. All of this will become clearer as we proceed to take up particular equations and particular loci.

Since it requires an ordered pair of numbers to define a point in the plane and any point in the plane is so defined, we say that a plane is *two-dimensional.* A line is *one-dimensional,* since any point on the line is defined by one number, for example, its distance from a fixed point on the line. Similarly, since any point on a curve is defined by one number, for example, its distance measured along the curve (say, with a piece of string) from a point of the curve, we say that any curve is one-dimensional. The graph of an equation in x and y is some kind of curve, or a line; for, as remarked above, only one of the unknowns may be chosen arbitrarily, the other being determined by the equation, and thus the equation picks out from the two-dimensional set of points in the plane a one-dimensional set.

We consider now the graph of an equation of the first degree in x and y. When one takes such an equation, and, having obtained a number of solutions after the manner discussed in § 1, plots their graphs, one observes that they seem to lie on a straight line. In fact, one may have been told that all one has to do is to plot two solutions and draw a straight line through the two points, and that this is the graph of the equation in the sense that the graph of every solution is a point of the line and every point of the line is the graph of a solution. In order to prove that this is a correct statement, we consider first the inverse problem of finding an equation of a line.

Consider a line which is not parallel to either axis, and choose upon it two particular points, P_1 of coordinates x_1, y_1 and P_2 of coordinates x_2, y_2. We denote by P of coordinates x, y any point on the line — meaning that it may be placed anywhere on the line while keeping P_1 and P_2 fixed; in this sense we speak of P as a *general* or *representative* point, and use x and y without subscripts to distinguish a representative point from a particular point. Through P_1, P, P_2 we draw lines parallel to the axes, forming similar triangles, as shown in Fig. 6. Since the triangles are similar, corresponding sides are proportional, and we have

$$\frac{P_1Q}{P_1Q_2} = \frac{QP}{Q_2P_2}. \tag{5.1}$$

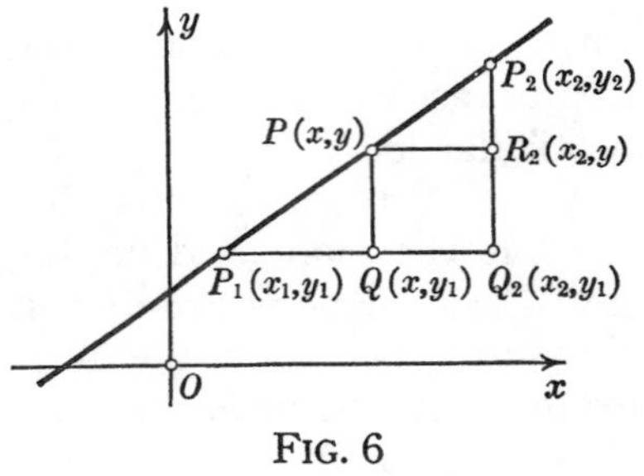

FIG. 6

It should be observed that the two segments in each ratio are taken in the same sense. We have $P_1Q = x - x_1$, $P_1Q_2 = x_2 - x_1$, $QP = y - y_1$, $Q_2P_2 = y_2 - y_1$. (Although Fig. 6 has been drawn with all the coordinates positive, the reader will readily verify that these relations hold equally well when the figure is so placed with reference to the axes that some or all of the numbers x_1, y_1, x, y, x_2, y_2 are negative.) Consequently (5.1) is equivalent to the equation

$$\frac{x - x_1}{x_2 - x_1} = \frac{y - y_1}{y_2 - y_1}. \tag{5.2}$$

Although this result has been derived for the case when P lies on the segment P_1P_2, by drawing suitable figures the reader can assure himself that equation (5.2) holds when P lies on the line but outside the segment P_1P_2. Equation (5.2) is reducible to

$$(y_2 - y_1)x - (x_2 - x_1)y + (x_2y_1 - x_1y_2) = 0. \tag{5.3}$$

This equation is satisfied by the coordinates of every point on the line. Moreover, from the form (5.2) of the equation it follows that any solution of the equation leads back to the ratios (5.1), and consequently gives the coordinates of a point on the line.

Although equation (5.3) was derived for a line inclined to the coordinate axes, that is, when $x_2 \neq x_1$ and $y_2 \neq y_1$, it applies to the cases when the line is parallel to either axis. In fact, if the line is parallel to the x-axis, we have $y_2 = y_1$, $x_2 \neq x_1$, in which case equation (5.3) is equivalent to

$$(5.4) \qquad y - y_1 = 0.$$

Also, when $x_2 = x_1$, $y_2 \neq y_1$, equation (5.3) is equivalent to

$$(5.5) \qquad x - x_1 = 0,$$

which is an equation of a line parallel to the y-axis. These results follow also from (5.2), if we adopt the principle that *when in an equation of two ratios either term of a ratio is equal to zero, the other term also is equal to zero.* Thus if $y_2 = y_1$, we must have $y - y_1 = 0$. This does not say anything about the value of the other ratio; this value is determined by the values of the terms of this ratio. Hence we have the theorem

[5.1] *Equation* (5.2), *or* (5.3), *is an equation of the line through the points* (x_1, y_1) *and* (x_2, y_2).

Since x_1, y_1, x_2, and y_2 are fixed numbers, equations (5.2) and (5.3) are of the first degree in x and y, as are also (5.4) and (5.5). Consequently we have the theorem

[5.2] *An equation of a straight line is of the first degree in x and y.*

We say "an equation," and not "the equation," because, if we have an equation of a line, so also is any constant multiple of this equation an equation of the line, since any solution of either equation is a solution of the other.

Having shown that in an equation of any line the coefficients of x and y are not both simultaneously equal to zero, we remark that if, as stated in § 1, in deriving theorems in this chapter we exclude from consideration the degenerate case of equations of the first degree for which the coefficients of both x and y are zero, we are not thereby restricting the consideration of all the lines in the plane.

We shall prove the converse of Theorem [**5.2**], namely,

[**5.3**] *The graph of any equation of the first degree in x and y is a straight line.*

We consider the general equation of the first degree

$$ax + by + c = 0 \tag{5.6}$$

in which not both a and b are equal to zero. For $a = 0$, $b \neq 0$, equation (5.6) is reducible to the form (5.4) (with a change of notation), which is an equation of a line parallel to the x-axis. For $a \neq 0$, $b = 0$, equation (5.6) is reducible to the form (5.5), which is an equation of a line parallel to the y-axis. In § 1 it was shown that when $a \neq 0$, $b \neq 0$, equation (5.6) can be put in the form (1.8) in terms of two of its solutions x_1, y_1 and x_2, y_2. But equation (1.8) is the same as equation (5.3). Hence any equation (5.6) can be given one of the forms (5.3), (5.4), or (5.5), and the theorem is proved.

Returning to the consideration of equation (5.2), we remark that, when the line is not parallel to either axis, for each point on the line the two ratios have the same value, this value, say t, depending upon the values of x and y. If we put each of the ratios equal to t and solve the resulting equations for x and y, we obtain

$$x = x_1 + t(x_2 - x_1), \qquad y = y_1 + t(y_2 - y_1). \tag{5.7}$$

Conversely, for each value of t the values of x and y given by (5.7) are such that equation (5.2) is satisfied, as is seen by substitution, and consequently these values of x and y are the coordinates of a point on the line. When the line is parallel to the x-axis, equations (5.7) hold, the second equation reducing to $y = y_1$; and similarly for a line parallel to the y-axis. Thus the single equation (5.2) of the line has been replaced by the two equations (5.7) through the introduction of an auxiliary variable t, called a *parameter*. Accordingly equations (5.7) are called *parametric equations* of the line. Ordinarily, as will be seen later, a curve in the plane is defined by one equation in x and y, but sometimes it is convenient to define it by two equations in x, y, and a parameter; an example of this is found in mechanics when the coordinates of a moving particle are expressed in terms of time as parameter. When the equations of a curve are in parametric form, it may be possible to obtain a single equation in x and y by eliminating the parameter from the two equations. Thus, if we solve the first of (5.7) for t and substitute the result in the second, we get an equation reducible to (5.2). Again, consider the locus of a point whose coordinates are given by $x = t^2$, $y = 2t$, as t takes on all values. Eliminating t, we get $y^2 = 4x$, which the reader who has plotted curves in his study of algebra will identify as an equation of a parabola.

The use of a parameter in defining a line, or a curve, emphasizes the fact that a line, or a curve, is one-dimensional in that any point on it is specified by the appropriate value of a single variable, a parameter. In fact, in plotting the graph of a line, or a curve, defined by parametric equations one assigns different values to the parameter and plots the points having as coordinates the values of x and y thus obtained; it is not necessary first to eliminate the parameter from the two equations and then plot the resulting equation in x and y (see Ex. 11).

When equations (5.7) are written in the form

$$(5.8) \qquad x = (1 - t)x_1 + tx_2, \qquad y = (1 - t)y_1 + ty_2,$$

we see that y is equal to the sum of the same multiples of y_1 and y_2 as x is of x_1 and x_2. We say that x and y have the same *linear* and *homogeneous* expressions in x_1, x_2 and y_1, y_2 respectively, meaning that every term in each expression is of the first degree in these quantities. Accordingly we have

[5.4] *Any point of a line is expressible linearly and homogeneously in terms of two fixed points of the line.*

Any two points of the line can be used for P_1 and P_2, that is, as the basis for writing the equations of the line in the form (5.8). The above theorem is the geometric equivalent of the algebraic statement

[5.5] *Any solution of an equation $ax + by + c = 0$ is expressible linearly and homogeneously in terms of any two solutions.*

EXERCISES

1. Obtain equations of the lines determined by the points:

a. $(-1, 8)$ and $(4, -2)$. *b.* $(3, 2)$ and $(3, -1)$.

2. Show that the points $(4, -3)$, $(2, 0)$, $(-2, 6)$ lie on a line.

3. Find c so that the points $(c, -2)$, $(3, 1)$, and $(-2, 4)$ shall be collinear, that is, lie on a line.

4. Where are the points for which $x > 0$, $y > 0$, and $x + y < 1$? Where are the points for which $2x - 3y > 2$?

5. Show that the equation of the second and third ratios in (4.3) is reducible to equation (5.3). What does this mean?

6. Show that (4.8) are of the form (5.8) with $t = h_1/(h_1 + h_2)$.

7. In what ratio is the line segment with end points $(2, 3)$ and $(3, -1)$ divided by the point of intersection of the line segment and the line $3x - 2y - 6 = 0$?

8. Show that the equation $a(x - x_1) + b(y - y_1) = 0$ for each set of values of a and b is an equation of a line through the point (x_1, y_1). For what values of a and b is it an equation of a line parallel to the y-axis; an equation of the line through the point (x_2, y_2)?

9. Show that the line $ax + by + c = 0$ meets the x-axis and y-axis in the points $P_1(-c/a, 0)$ and $P_2(0, -c/b)$ respectively; when $c \neq 0$, the lengths OP_1 and OP_2, that is, $-c/a$ and $-c/b$, are called the *x-intercept* and *y-intercept* respectively of the line.

10. Show that when an equation of a line is written in the form

$$\frac{x}{g} + \frac{y}{h} = 1,$$

where g and h are constants, g and h are the x- and y-intercepts respectively. When can an equation of a line not be given this form, which is called *the equation of the line in the intercept form*?

11. Plot the curve with the parametric equations $x = 2t^2$, $y = 3t$; also the curve $x = 5 \cos t$, $y = 5 \sin t$, using a table of natural sines and cosines. What is the equation of the curve in each case when the parameter is eliminated?

12. Show that, if the expressions (4.8) for x and y are substituted in an equation $ax + by + c = 0$, the value of h_1/h_2 obtained from the result is the ratio in which the segment P_1P_2 is divided by the point of intersection of the line $ax + by + c = 0$ and the line through P_1 and P_2. The reader should apply this method to Ex. 7.

6. Direction Numbers and Direction Cosines of a Line. Angle between Two Lines

The equation

$$\frac{x - x_1}{x_2 - x_1} = \frac{y - y_1}{y_2 - y_1} \tag{6.1}$$

of the line through the points (x_1, y_1) and (x_2, y_2), obtained in § 5, expresses the condition that the direction numbers of the segments P_1P and P_1P_2 are proportional when P is any point on the line. In fact, in Fig. 6 the sensed segments P_1Q and QP give the direction numbers of P_1P, and P_1Q_2 and Q_2P_2 the direction numbers of P_1P_2. Thus equation (6.1) is the algebraic statement of the characteristic property of a line that any two segments having an end point in common have the same direction. If we say that direction numbers of any segment of a line are *direction numbers of the line,* it follows that a line has an endless number of sets of direction numbers, but that the numbers of any set are proportional to those of any other set.

The equation

$$\frac{x - x_1}{u} = \frac{y - y_1}{v}, \tag{6.2}$$

which is satisfied by x_1 and y_1, being of the first degree in x and y, is an equation of a line through the point (x_1, y_1) and with direction numbers u and v. In order to find a segment of which u and v are direction numbers, we have only to find x_2 and y_2 from the equations

$$x_2 - x_1 = u, \qquad y_2 - y_1 = v.$$

Then P_1P_2 is a segment with end points $P_1(x_1, y_1)$ and $P_2(x_2, y_2)$ having u and v as direction numbers; and in terms of x_2 and y_2 equation (6.2) becomes equation (6.1). Since *parallel lines* have the same direction by definition, it follows that

[6.1] *If (x_1, y_1) and (x_2, y_2) are any two points on a line, $x_2 - x_1$ and $y_2 - y_1$ are direction numbers of the line, and also of any line parallel to it.*

In defining the direction cosines of a line segment in § 3 we assigned sense to a segment, making the convention that for a segment not parallel to the x-axis upward along the segment is positive and downward negative; and for a segment parallel to the x-axis the positive sense is to the right along the segment. Accordingly the positive sense along a line is upward, that is, y increasing, when the line is not parallel to the x-axis; and to the right, that is, x increasing, when the line is parallel to the x-axis. Since all segments of a line have the same direction, the direction cosines of all segments are the same; we call them *the direction cosines of the line.* Accordingly the *direction cosines* λ, μ *of a line are the cosines of the angles between the positive direction of the line and the positive directions of the x- and y-axes respectively.*

From Theorem [**3.2**] we have

[**6.2**] *The direction cosines* λ, μ *of a line satisfy the equation*

$$\lambda^2 + \mu^2 = 1; \tag{6.3}$$

μ *is never negative; when* $\mu > 0$, $-1 < \lambda < +1$; *when* $\mu = 0$, $\lambda = 1$; *and, conversely, any two numbers* λ, μ *satisfying these conditions are direction cosines of a line.*

In Figs. 7 and 8

$$\lambda = \cos A, \quad \mu = \cos B. \tag{6.4}$$

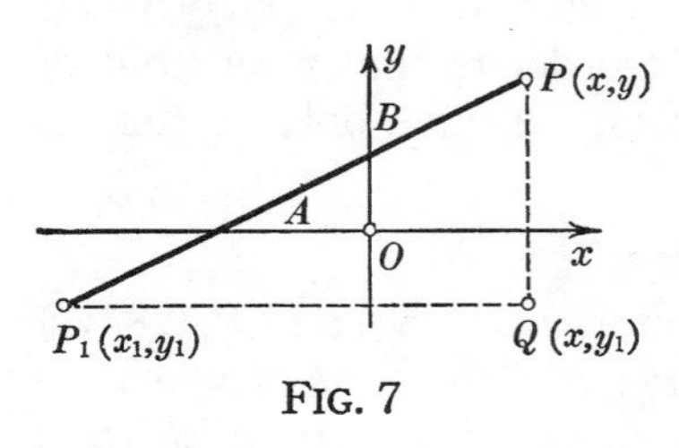

FIG. 7

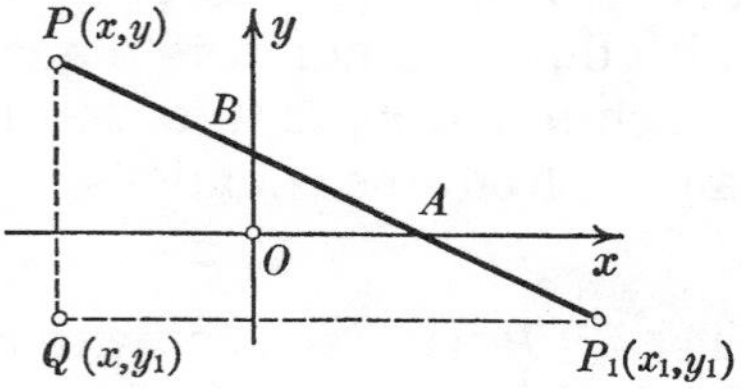

FIG. 8

Since equations (3.5) hold for all segments of a line with the point (x_1, y_1) as an end point of the segment, and consequently for the line, we have

[**6.3**] *The equations*

$$x = x_1 + d\lambda, \quad y = y_1 + d\mu \tag{6.5}$$

are parametric equations of the line through the point (x_1, y_1)

and with direction cosines λ and μ; the parameter d is the distance from the point (x_1, y_1) to a representative point (x, y), and is positive or negative according as the latter point is above or below P_1 along the line, or to the right or left of P_1 when the line is parallel to the x-axis.

With the understanding that for any quantity A the symbol $\sqrt{A}$ means the positive square root of A, and that $\sqrt{a^2} = |a|$, as defined in § 4, we shall prove the theorem

[6.4] *The direction cosines of a line for which u and v are any set of direction numbers are given by*

$$\lambda = \frac{u}{e\sqrt{u^2 + v^2}}, \qquad \mu = \frac{v}{e\sqrt{u^2 + v^2}}, \tag{6.6}$$

where e is $+1$ or -1 so that ev is positive when $v \neq 0$, and eu is positive when $v = 0$.

In fact, the quantities (6.6) satisfy the condition (6.3) of Theorem [**6.2**]; and the requirement in Theorem [**6.4**] concerning e is such that μ is positive when not equal to zero, and that when $\mu = 0$, then $\lambda = 1$, which proves the theorem.

From Theorem [**3.3**], using the Greek letter ϕ (phi), we have

[6.5] *The angle ϕ between the positive directions of two lines whose direction cosines are λ_1, μ_1 and λ_2, μ_2 is given by*

$$\cos\phi = \lambda_1\lambda_2 + \mu_1\mu_2. \tag{6.7}$$

From Theorems [**6.5**] and [**6.4**] we have

[6.6] *The angle ϕ between the positive directions of two lines whose direction numbers are u_1, v_1 and u_2, v_2 is given by*

$$\cos\phi = \frac{u_1u_2 + v_1v_2}{e_1e_2\sqrt{(u_1^2 + v_1^2)(u_2^2 + v_2^2)}}, \tag{6.8}$$

where e_1 is $+1$ or -1 so that e_1v_1 is positive when $v_1 \neq 0$, and e_1u_1 is positive when $v_1 = 0$; and similarly for e_2.

As a corollary of the theorem we have

[6.7] *Two lines with direction numbers u_1, v_1 and u_2, v_2 are perpendicular to one another, if and only if*

$$u_1u_2 + v_1v_2 = 0. \tag{6.9}$$

Consider now an equation

$$ax + by + c = 0 \tag{6.10}$$

of a line. If the point (x_1, y_1) is on the line, c must be such that

$$ax_1 + by_1 + c = 0.$$

Subtracting this equation from (6.10), we have as an equation of the line

$$a(x - x_1) + b(y - y_1) = 0. \tag{6.11}$$

When a and b are different from zero, and this equation is written in the form

$$\frac{x - x_1}{b} = \frac{y - y_1}{-a},$$

we see that b and $-a$ are direction numbers of the line, and consequently of any line parallel to it. Hence we have

[6.8] *An equation of any one of the endless number of lines parallel to the line $ax + by + c = 0$ is*

$$ax + by + d = 0; \tag{6.12}$$

a particular one of these lines is determined by the value of the coefficient d.

For example, if we wish to find an equation of the line through the point $(3, -2)$ parallel to the line $4x - y + 5 = 0$, d is determined by $4(3) - (-2) + d = 0$, that is, $d = -14$, and hence an equation of the desired line is $4x - y - 14 = 0$.

We seek next an equation of any line perpendicular to the line with equation (6.10). If we denote by u and v direction numbers of such a perpendicular, and note that b, $-a$ are direction numbers of the line (6.10), it follows from Theorem [6.7] that u and v must be such that

$$bu - av = 0.$$

This condition is satisfied by $u = a$, $v = b$, and by any constant multiple of a and b. Hence we have

[6.9] *The geometric significance of the coefficients a and b of an equation $ax + by + c = 0$ of a line is that a and b are direction numbers of any line perpendicular to the given line.*

In particular, an equation of the line perpendicular to the line with equation (6.11) at the point (x_1, y_1) of the latter is

$$\frac{x - x_1}{a} = \frac{y - y_1}{b}.$$

This result, illustrated in Fig. 9, may be stated as follows:

[6.10] *Either of the lines*

$$a(x - x_1) + b(y - y_1) = 0, \qquad \frac{x - x_1}{a} = \frac{y - y_1}{b} \tag{6.13}$$

is the perpendicular to the other at the point (x_1, y_1).

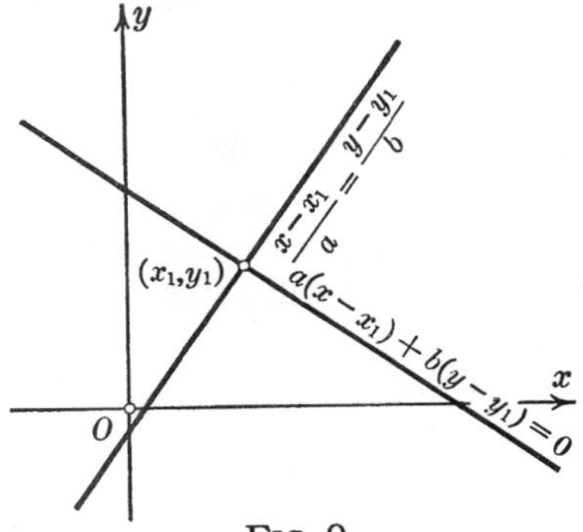

FIG. 9

When the second of equations (6.13) is written in the form

$$b(x - x_1) - a(y - y_1) = 0, \tag{6.14}$$

we see that the coefficients of x and y are b and $-a$ respectively. Hence we have

[6.11] *Any line perpendicular to the line with the equation $ax + by + c = 0$ has for an equation*

$$bx - ay + d = 0; \tag{6.15}$$

a particular line is determined by the value of d.

EXERCISES

1. Find an equation of the line through the point (3, 0) and parallel to the line $3x - y + 5 = 0$.

2. What are the direction cosines of the line $3x - 4y + 1 = 0$; of a line perpendicular to it?

3. Find an equation of all lines perpendicular to $2x + 5y - 2 = 0$, and in particular of the one passing through the point (1, − 1).

4. Find equations of the lines through the intersection of $x - y + 2 = 0$ and $4x - y - 1 = 0$ parallel and perpendicular respectively to the line $2x + 5y - 3 = 0$.

5. Find the angle between the positive directions of the perpendiculars to each of the following pairs of lines:

a. $2x - 7y + 3 = 0, \quad 5x + y + 1 = 0.$
b. $3x - y - 10 = 0, \quad 3x - y + 8 = 0.$
c. $5x - 2y + 1 = 0, \quad 2x + 5y - 1 = 0.$

6. Find the angles between the positive directions of the lines in Ex. 5.

7. Show that the coordinates of any point on a line through the origin are direction numbers of the line.

8. Show that equations (6.5) hold for a line parallel to either coordinate axis if we make the convention that "the angle between" the positive directions of two parallel lines is zero.

9. Show that for the lines in Figs. 7 and 8 the equation (6.3) is equivalent to the trigonometric identity $\sin^2 A + \cos^2 A = 1$.

10. Show that

$$\frac{x - x_1}{x_2 - x_3} = \frac{y - y_1}{y_2 - y_3}, \qquad (x - x_1)(x_2 - x_3) + (y - y_1)(y_2 - y_3) = 0$$

are equations of two lines through the point (x_1, y_1) parallel and perpendicular respectively to the line through the points (x_2, y_2) and (x_3, y_3).

11. Given the triangle whose vertices are $A(a, 0)$, $B(0, b)$, and $C(c, 0)$; prove that the perpendiculars from the vertices of the triangle upon the opposite sides meet in a point; that the perpendicular bisectors of the sides meet in a point. Does this prove that these results are true for any triangle?

12. Discuss equation (6.2) and Theorem [**6.4**] when one of the direction numbers is zero.

7. The Slope of a Line

By definition the *slope* of a line, which is usually denoted by m, is the tangent of the angle which the positive direction of the line, as defined in § 6, makes with the positive direction of the x-axis; it is the tangent of the angle measured from the x-axis to the line in the counterclockwise direction (as is done in trigonometry). In Fig. 7 the angle at P_1 is equal to A, and consequently

$$m = \tan A = \frac{QP}{P_1Q} = \frac{y - y_1}{x - x_1}. \tag{7.1}$$

In Fig. 8 the angle at P_1 is the supplement of A; consequently

$$\tan (180° - A) = \frac{QP}{QP_1} = \frac{y - y_1}{x_1 - x},$$

which is equivalent to (7.1), since $\tan (180° - A) = -\tan A$. It is customary to write (7.1) in the form

$$y - y_1 = m(x - x_1), \tag{7.2}$$

as an equation of the line expressed in terms of the slope and a point (x_1, y_1) on the line.

When the line is parallel to the y-axis, the denominator in the right-hand member of (7.1) is zero. Hence we say that the slope is not defined for a line parallel to the y-axis; some writers say that in this case the slope is infinite.

In equation (7.1) the quantities $x - x_1$ and $y - y_1$ are direction numbers of the line segment P_1P and consequently of the line; since $P(x, y)$ and $P_1(x_1, y_1)$ are any points of the line, we have

[7.1] *The slope m of a line not parallel to the y-axis is equal to the ratio of the second and first of any set of direction numbers of the line; in particular, $m = (y_2 - y_1)/(x_2 - x_1)$, where (x_1, y_1) and (x_2, y_2) are any two points of the line.*

Equation (7.2) is a particular case of the equation

$$y = mx + h, \tag{7.3}$$

where $h = y_1 - mx_1$. For a given value of m equation (7.3) is an equation of all lines with the slope m, that is, a set of parallel lines, any line of the set being determined by a suitable value of h. In particular, equation (7.2) is an equation of the line of the set through the point (x_1, y_1). An equation

$$ax + by + c = 0, \tag{7.4}$$

for which $b \neq 0$, can be put in the form (7.3) by solving the given equation for y. Hence we have

[7.2] *If an equation $ax + by + c = 0$ of a line not parallel to the y-axis is solved for y, the coefficient of x in the resulting equation is the slope of the line.*

As a consequence of theorems [**7.1**] and [**6.7**] we have

[7.3] *Two lines with the slopes m_1 and m_2 are perpendicular to one another, if and only if*

$$m_1 m_2 + 1 = 0; \tag{7.5}$$

that is, if either slope is the negative reciprocal of the other.

The proof is left to the reader.

We have introduced the concept of slope because of its traditional use, and particularly because of its use in the application of the differential calculus to the study of lines and curves in the plane, an interesting subject awaiting the reader. On the other hand, we have emphasized direction numbers and direction cosines because they may be applied to any line without exception, and, as we shall see in Chapter 2, are applicable to the study of lines in space, whereas the concept of slope is not used in this connection.

EXERCISES

1. Find an equation of all lines parallel to $y = 3x - 5$, and in particular the one through the point $(1, -3)$.

2. Find an equation of the line through the origin whose slope is twice that of the line $3x - y + 2 = 0$.

3. Find an equation of the line through the point $(2, -3)$ whose positive direction makes the angle of 60° with the positive x-axis.

4. Find equations of the two lines through the point $(-1, 3)$ forming with the x-axis an isosceles triangle with the point as vertex and the base angles 30°

5. For what value of m is the line (7.2) perpendicular to the line joining the point (x_1, y_1) to the origin?

6. Using the formula from trigonometry for the tangent of the difference of two angles, show that an angle θ between two lines whose slopes are m_1 and m_2 is given by

$$\tan \theta = \frac{m_2 - m_1}{1 + m_1 m_2}.$$

7. How is Theorem [**7.3**] a consequence of Ex. 6?

8. Find equations of the lines through the point (x_1, y_1) which make angles of 45° with the line (7.2), and verify that they are perpendicular to one another.

9. Show that the graph of the equation $ax^2 + 2\,hxy + by^2 = 0$ is a pair of straight lines through the origin if $h^2 > ab$. What are the slopes of these lines?

10. Draw the graph of a line and indicate the various quantities appearing in equations (6.1), (6.2), (6.5), and (7.2).

8. Directed Distance from a Line to a Point

When a point $P_1(x_1, y_1)$ is on the line

$$ax + by + c = 0, \tag{8.1}$$

the expression $ax_1 + by_1 + c$ is equal to zero, as we have seen. When P_1 is not on the line, this expression has a value different from zero. One might question whether perhaps this value has something to do with the distance from the line to the point. We shall answer this question by showing that this distance is $\dfrac{ax_1 + by_1 + c}{\sqrt{a^2 + b^2}}$, and give a meaning to the algebraic sign of the resulting number.

When $a = 0$ and $b \neq 0$ in (8.1), that is, when the line is parallel to the x-axis, the value of y for each point on the line is $-c/b$, and consequently the distance of P_1 from the line is $y_1 - (-c/b)$, that is, $y_1 + c/b$, and this distance is positive or negative according as P_1 lies above or below the line; we get this expression for the distance when we put $a = 0$ in the expression at the close of the preceding paragraph. When $b = 0$ and $a \neq 0$, that is, when the line is parallel to the y-axis, the distance is $x_1 + c/a$, which is positive or negative according as P_1 lies to the right or left of the line; this expression follows from the one of the preceding paragraph on putting $b = 0$.

We take up next the case when the line is inclined to the x-axis, in which case $a \neq 0$, $b \neq 0$. In § 6 we made the convention that the direction upward along a line inclined to the axes is positive, and showed that the direction cosine μ is always positive, and that λ is positive or negative according as the line is inclined as in Fig. 7 or in Fig. 8. In consequence of this result, and of Theorems [**6.9**] and [**6.4**], we have that the direction cosines of a line perpendicular to the line $ax + by + c = 0$ are given by

$$(8.2)\qquad \lambda = \frac{a}{e\sqrt{a^2 + b^2}}, \qquad \mu = \frac{b}{e\sqrt{a^2 + b^2}},$$

where e is $+1$ or -1 so that $eb > 0$ when $b \neq 0$, and $ea > 0$ when $b = 0$.

We consider first the case when the line (8.1) is inclined as in Fig. 10, where $P_1(x_1, y_1)$ is a point above the line and $P_2(x_2, y_2)$ is the point in which the perpendicular through P_1 to the line meets it. We denote by d the distance from P_2 to P_1, d being a positive number, since it is measured in the positive direction of the line P_2P_1. Comparing this line in Fig. 10, whose direction cosines are λ and μ, with the line P_1P in Fig. 7, we see that in place of (6.5) we have

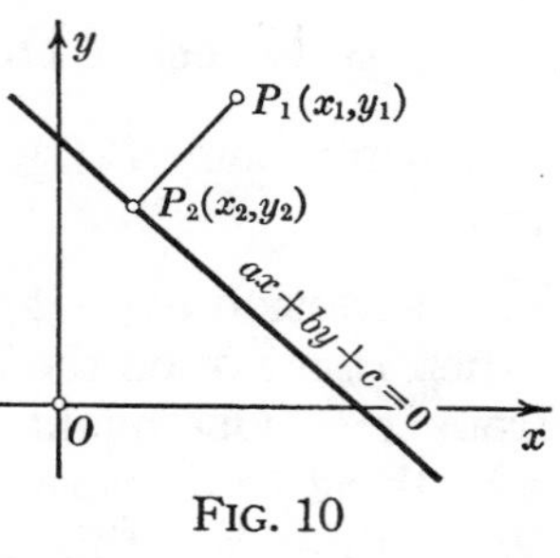

FIG. 10

$$(8.3)\qquad x_1 - x_2 = d\lambda, \qquad y_1 - y_2 = d\mu,$$

from which and (8.2) we have

$$(8.4)\qquad x_2 = x_1 - \frac{da}{e\sqrt{a^2 + b^2}}, \qquad y_2 = y_1 - \frac{db}{e\sqrt{a^2 + b^2}}.$$

Since (x_2, y_2) is a point on the line (8.1), on substituting the expressions (8.4) in (8.1) and rearranging terms, we have

$$ax_1 + by_1 + c - d\left(\frac{a^2 + b^2}{e\sqrt{a^2 + b^2}}\right) = 0,$$

from which, on solving for d and noting that $1/e = e$, we obtain

$$d = \frac{ax_1 + by_1 + c}{e\sqrt{a^2 + b^2}}.$$

When P_1 lies below the line, equations (8.3) hold equally well provided d is a negative number, as the reader may verify by drawing a figure and noting that when the line (8.1) is inclined as in Fig. 10, λ is positive.

We consider the other case when the line (8.1) is inclined as in Fig. 11, and this time take P_1 below the line, so as to add variety to the discussion. Comparing the line P_1P_2 in Fig. 11 with the line P_1P in Fig. 8, we have from (6.5)

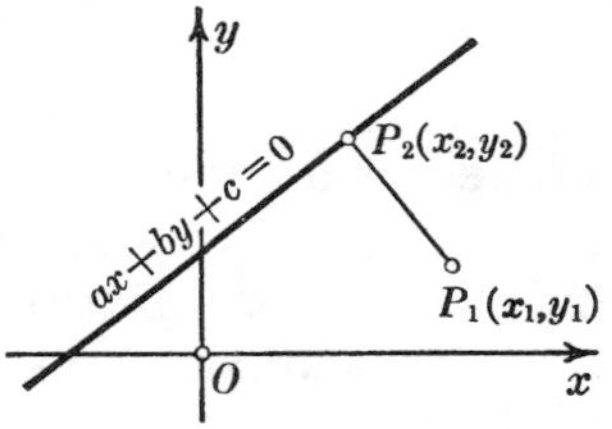

FIG. 11

$$x_2 - x_1 = d\lambda, \qquad y_2 - y_1 = d\mu,$$

where d is a positive number, the directed distance P_1P_2. Comparing these equations with (8.3), we see that the left-hand members of these respective equations differ only in sign. Consequently equations (8.3) apply in this case also, with the understanding that d is negative; it is the directed distance P_2P_1, that is, P_1 is on the negative side of the line. The reader may verify that equations (8.3) hold also when P_1 is above the line in Fig. 11 and d is positive. Hence we have

[8.1] *The directed distance of a point $P_1(x_1, y_1)$ from the line $ax + by + c = 0$ is given by*

$$d = \frac{ax_1 + by_1 + c}{e\sqrt{a^2 + b^2}}, \tag{8.5}$$

where e is $+1$ or -1 so that $eb > 0$ if $b \neq 0$, and $ea > 0$ if $b = 0$; the distance d is positive or negative according as the point lies above or below the line.

As an example, what is the distance of the point $(-2, 1)$ from the line $x - 3y + 2 = 0$? Since the coefficient of y is a negative number, $e = -1$ in (8.5), and we have

$$d = \frac{(-2) - 3(1) + 2}{-\sqrt{10}} = \frac{3}{\sqrt{10}}.$$

Since d is positive, the point lies above the line, as the reader may verify by drawing the graph.

We observe from (8.5) that the origin, that is, the point $(0, 0)$, is above or below the line according as ec is positive or negative.

Consider now the equation

$$\frac{ax + by + c}{e\sqrt{a^2 + b^2}} = k, \tag{8.6}$$

where k is a constant. Since the equation is of the first degree in x and y, it is an equation of a line. If $P_1(x_1, y_1)$ is any point on this line, on substituting x_1 and y_1 for x and y in the above equation and comparing the result with (8.5), we see that P_1 is at the distance k from the line (8.1). Since this is true of every point on the line (8.6), it follows that the line (8.6) is parallel to the line (8.1) and at the directed distance k from it; that the two lines are parallel follows also from the fact that equations (8.1) and (8.6) satisfy the condition of theorem [**6.8**].

EXERCISES

1. Find the distances from the line $x - y + 5 = 0$ to the points $(1, -1)$ and $(-2, 4)$, and find equations of the lines through these points parallel to the given line.

2. Find equations of the two lines parallel to $4x - 3y + 5 = 0$ at the distance 2 from this line.

3. Find the distance between the lines $2x - y + 6 = 0$ and $4x - 2y - 3 = 0$.

4. Find the points on the x-axis whose distances from the line $4x + 3y - 6 = 0$ are numerically equal to 3.

5. How far is the origin from the line through the point $(-3, 2)$ parallel to the line $3x + y = 0$?

6. Find the points which are equidistant from the points (2, 6) and (– 2, 5) and at a distance of 2 units from the line $7x + 24y - 50 = 0$.

7. Show that the circle of radius 2 and center (3, 4) is not intersected by the line $4x - 7y + 28 = 0$.

8. Find the area of the triangle with the vertices (2, 3), (5, – 1), (3, 1).

9. By what factor must the equation $ax + by + c = 0$ be multiplied so that it is of the form $\lambda x + \mu y + p = 0$, where λ and μ are the direction cosines of any line perpendicular to the given line? Show that p is the directed distance from the line to the origin.

10. Prove that the line $2x - 4y + 8 = 0$ bisects the portion of the plane between the two lines $x - 2y + 2 = 0$ and $x - 2y + 6 = 0$.

11. Prove that the line $3x + 4y - 12 = 0$ bisects the area of the quadrilateral whose vertices are $(-4, 6)$, $(7, \frac{3}{2})$, $(1, -\frac{3}{2})$, $(8, -3)$.

9. Two Equations of the First Degree in x and y. Determinants of the Second Order

Having discussed in § 1 the solutions of an equation of the first degree in x and y, we now consider two such equations, namely,

$$(9.1) \qquad a_1x + b_1y + c_1 = 0, \qquad a_2x + b_2y + c_2 = 0,$$

where a subscript 1 attached to a coefficient (as, for example, a_1) indicates that it is a coefficient of the first equation and has nothing to do with the numerical value of the coefficient; similarly a_2, b_2, c_2 are coefficients of the second equation; this device of subscripts enables one to use the same letter for the coefficients of x in the two equations, the distinction being indicated by the subscripts; similarly for the coefficients of y and the constant terms. In accordance with the discussion in § 1, each of equations (9.1) has an endless number of solutions; x and y in the first equation represent *any of its solutions*, and in the second x and y represent *any of its solutions*. When the left-hand member of either equation is a constant multiple of the left-hand member of the other, the two equations have the same solutions, as remarked in § 1; that is, they are not independent equations, and consequently x and y have the same meaning in both equations. But when the equations are essentially different, x and y have, in general, different meanings in the two equations. To emphasize this fact, different letters might more appropriately be used in place of x and y in the second equation, but this is not the general practice; we adhere to the customary practice of using the same letters x and y in both equations and expect the reader to bear the distinction in mind.

It may or may not be that for given values of the coefficients there is a solution of the first which is also a solution of the second; that is, the two equations may or may not have a *common* solution. We assume that there is a common solution, which we denote by x_1, y_1, and substitute it for x and y in the two equations, obtaining

$$(9.2) \qquad a_1x_1 + b_1y_1 + c_1 = 0, \qquad a_2x_1 + b_2y_1 + c_2 = 0.$$

Following the method with which the reader is familiar, we multiply the first of these equations by b_2 and from the result subtract the second multiplied by b_1; in the resulting equation the coefficient of y_1 is zero, and we obtain

$$(a_1b_2 - a_2b_1)x_1 + (c_1b_2 - c_2b_1) = 0.$$

By adopting the following shorthand notation for the quantities in parentheses:

$$(9.3) \qquad \begin{vmatrix} a_1 & b_1 \\ a_2 & b_2 \end{vmatrix} \equiv a_1b_2 - a_2b_1, \qquad \begin{vmatrix} c_1 & b_1 \\ c_2 & b_2 \end{vmatrix} \equiv c_1b_2 - c_2b_1,$$

the above equation may be written

$$(9.4) \qquad \begin{vmatrix} a_1 & b_1 \\ a_2 & b_2 \end{vmatrix} x_1 + \begin{vmatrix} c_1 & b_1 \\ c_2 & b_2 \end{vmatrix} = 0.$$

In like manner, if we multiply the second of equations (9.2) by a_1 and from the result subtract the first multiplied by a_2, we obtain

$$(9.5) \qquad \begin{vmatrix} a_1 & b_1 \\ a_2 & b_2 \end{vmatrix} y_1 + \begin{vmatrix} a_1 & c_1 \\ a_2 & c_2 \end{vmatrix} = 0,$$

where, similarly to (9.3), we have put

$$\begin{vmatrix} a_1 & c_1 \\ a_2 & c_2 \end{vmatrix} \equiv a_1c_2 - a_2c_1. \tag{9.6}$$

The square arrays defined by (9.3) and (9.6) are called *determinants.* We use the sign of identity $\equiv$, rather than the sign of equality, to indicate that (9.3) and (9.6) are definitions.

We observe that the second determinants in (9.4) and in (9.5) are obtained from the first on replacing the a's by c's and the b's by c's respectively.

We consider equations (9.4) and (9.5) in detail. Suppose first that the a's and b's have such values that the determinant $\begin{vmatrix} a_1 & b_1 \\ a_2 & b_2 \end{vmatrix}$ is not equal to zero. Then equations (9.4) and (9.5) can be solved at once for x_1 and y_1, the common solution; this is the process with which the reader is familiar, although maybe not in this notation. Moreover, since any common solution of equations (9.1) must be a solution of equations (9.4) and (9.5), there is only one common solution in this case. We call the above determinant, that is, the determinant of the coefficients of x and y in equations (9.1), *the determinant of the equations.* Accordingly we have

[9.1] *When the determinant of two equations of the first degree in x and y is not equal to zero, there is one and only one common solution of the equations.*

It is evident that two lines with equations (9.1) meet in one point and only one point, that is, the lines intersect one another, if and only if their equations have one and only one common solution, in which case the point of intersection is the graph of the common solution. Accordingly the geometric equivalent of Theorem [**9.1**] is the following:

[9.2] *The lines with equations* (9.1) *intersect one another, that is, have one and only one point in common, if*

$$\begin{vmatrix} a_1 & b_1 \\ a_2 & b_2 \end{vmatrix} \neq 0. \tag{9.7}$$

We consider next the case when the determinant of the equations (9.1) is equal to zero, that is,

$$a_1b_2 - a_2b_1 = 0. \tag{9.8}$$

If at the same time $\begin{vmatrix} c_1 & b_1 \\ c_2 & b_2 \end{vmatrix} \neq 0$ or $\begin{vmatrix} a_1 & c_1 \\ a_2 & c_2 \end{vmatrix} \neq 0$, it follows that (9.4) or (9.5) cannot be true. This means that the assumption that there is a common solution of equations (9.1) is not valid, not that one has proved that zero is equal to a number which is not zero. Hence we have the theorem

[9.3] *Two equations* (9.1) *do not have a common solution if the determinant of the equations is equal to zero and one of the determinants*

$$\begin{vmatrix} c_1 & b_1 \\ c_2 & b_2 \end{vmatrix}, \qquad \begin{vmatrix} a_1 & c_1 \\ a_2 & c_2 \end{vmatrix} \tag{9.9}$$

is not equal to zero.

We shall show that in this case the equations have a form which is readily distinguishable, and to this end we consider separately the case when $a_1 \neq 0$ and the case when $a_1 = 0$.

Case 1. $a_1 \neq 0$. We cannot have $a_2 = 0$, otherwise it follows from (9.8) that $b_2 = 0$, and thus the second of equations (9.1) is of the degenerate type which we have excluded from our consideration (see § 1). Since $a_2 \neq 0$, a number k ($\neq 0$) is defined by $k = a_2/a_1$, from which we have $a_2 = ka_1$. When this value for a_2 is substituted in (9.8), we obtain the second of the following equations:

$$a_2 = ka_1, \qquad b_2 = kb_1. \tag{9.10}$$

Case 2. $a_1 = 0$. From (9.8) it follows that $a_2 = 0$, otherwise $b_1 = 0$, and the first of equations (9.1) is degenerate. Since $b_2 \neq 0$ for a similar reason, the second of equations (9.10) is satisfied by $k = b_2/b_1$, and the first of equations (9.10) is satisfied identically, since $a_1 = a_2 = 0$.

Thus equations (9.10) are a consequence of (9.8), and, conversely, when the coefficients of equations (9.1) are related as in (9.10) for some value of k, equation (9.8) holds, as is seen by substitution. Furthermore, when (9.10) holds and either of the determinants $\begin{vmatrix} c_1 & b_1 \\ c_2 & b_2 \end{vmatrix}$ and $\begin{vmatrix} a_1 & c_1 \\ a_2 & c_2 \end{vmatrix}$ is not equal to zero, we must have $c_2 \neq kc_1$. Accordingly we have

[9.4] *Two equations* (9.1) *are of the form*

$$(9.11) \qquad a_1x + b_1y + c_1 = 0, \qquad k(a_1x + b_1y) + c_2 = 0 \qquad (c_2 \neq kc_1),$$

where k is some constant different from zero, if and only if the determinant of the equations (9.1) *is equal to zero and one of the determinants* (9.9) *is not equal to zero; in this case there is no common solution of the equations.*

If we divide the second of equations (9.11) by k and put $c_2/k = d$, then $d \neq c_1$. If then we say that the second of (9.11) is $a_1x + b_1y + d = 0$ to within a constant factor, we may state Theorem [**9.4**] as follows:

[9.5] *Two equations of the first degree in x and y have no common solution if to within possible constant factors they are of the forms*

$$(9.12) \qquad ax + by + c = 0, \qquad ax + by + d = 0 \qquad (d \neq c).$$

In accordance with Theorem [**6.8**], two lines with equations (9.12) are parallel; that is, by the definition in § 6 they have the same direction. By Theorem [**9.5**] this definition is equivalent to the definition that they do not have a point in common. The reader may have been told that two parallel lines meet in a point at infinity. A point at infinity so defined is not like the points with which we are dealing; it is a concept sometimes introduced by the geometer so that he may make the general statement that any two noncoincident lines meet in a point.

We consider finally the case when all three of the determinants in (9.4) and (9.5) are equal to zero, that is, when we have equation (9.8) and

$$(9.13) \qquad c_1b_2 - c_2b_1 = 0, \qquad a_1c_2 - a_2c_1 = 0.$$

In this case any value of x_1 and any value of y_1 satisfy equations (9.4) and (9.5). This does not mean that any value of

x and any value of y is a common solution of equations (9.1), because we know from § 1 that any value of x and any value of y is not a solution of either equation. We shall show that it means that either equation is a constant multiple of the other.

We have seen that equation (9.8) is equivalent to the two equations (9.10). When the values for a_2 and b_2 from (9.10) are substituted in (9.13), we find that the latter are satisfied, if and only if $c_2 = kc_1$. From this result and (9.10) we have the theorem

[9.6] *All three of the determinants*

$$\begin{vmatrix} a_1 & b_1 \\ a_2 & b_2 \end{vmatrix}, \quad \begin{vmatrix} c_1 & b_1 \\ c_2 & b_2 \end{vmatrix}, \quad \begin{vmatrix} a_1 & c_1 \\ a_2 & c_2 \end{vmatrix} \tag{9.14}$$

are equal to zero, if and only if

$$\frac{a_2}{a_1} = \frac{b_2}{b_1} = \frac{c_2}{c_1}, \tag{9.15}$$

with the understanding that if either term in any ratio is equal to zero so also is the other term; for example, if $b_2 = 0$, *so also is* $b_1 = 0$.

This theorem is equivalent to the statement that all three of the determinants (9.14) are equal to zero, if and only if either of equations (9.1) is a constant multiple of the other, in which case any solution of either equation is a solution of the other. Hence we have the theorem

[9.7] *Equations* (9.1) *have an endless number of common solutions, if and only if all three of the determinants* (9.14) *are equal to zero.*

The geometric equivalent of theorem [**9.7**] is the following:

[9.8] *Two lines with the equations* (9.1) *are coincident, that is, coincide at every point, if and only if all three of the determinants* (9.14) *are equal to zero.*

The above discussion is set forth in the following table, the last column giving the number of common solutions of the two

equations (9.1), or, what is the geometric equivalent, the number of common points of two lines with these equations:

$\begin{vmatrix} a_1 & b_1 \\ a_2 & b_2 \end{vmatrix} \neq 0$				One
$\begin{vmatrix} a_1 & b_1 \\ a_2 & b_2 \end{vmatrix} = 0$	$\begin{vmatrix} c_1 & b_1 \\ c_2 & b_2 \end{vmatrix}$, $\begin{vmatrix} a_1 & c_1 \\ a_2 & c_2 \end{vmatrix}$ not both zero			None
	$\begin{vmatrix} c_1 & b_1 \\ c_2 & b_2 \end{vmatrix} = \begin{vmatrix} a_1 & c_1 \\ a_2 & c_2 \end{vmatrix} = 0$	a_1, b_1, a_2, b_2 not all zero		An endless number
		$a_1 = b_1 = a_2 = b_2 = 0$	c_1, c_2 not both zero	None
			$c_1 = c_2 = 0$	Any x and any y

The case $a_1 = b_1 = a_2 = b_2 = 0$ was not discussed in the text, that is, the case when both of equations (9.1) are degenerate, but the situation in this case is given in the table so that the algebraic treatment may be complete; the reader should verify the statements made.

In illustration of some of the above results we consider the following equations:

$$2x + 3y - 4 = 0, \qquad 4x - 5y + 3 = 0.$$

The determinant of these equations is $2(-5) - 4(3) = -22$. Hence, in accordance with Theorem [**9.1**], these equations admit one and only one common solution, given by equations (9.4) and (9.5), which in this case are

$$-22x_1 + 11 = 0, \qquad -22y_1 + 22 = 0,$$

that is, $x_1 = 1/2$, $y_1 = 1$.

If now we take the first of the above equations, $2x + 3y - 4 = 0$, and the equation $4x + 6y + 3 = 0$, we find that the determinant of these equations is equal to zero, and the second determinants of (9.4) and (9.5) are equal respectively to -33 and 22; consequently the given equations have no common solution, by Theorem [**9.3**].

When, however, we consider the equations $2x + 3y - 4 = 0$ and $4x + 6y - 8 = 0$, all three of the determinants in equations (9.4) and (9.5) are equal to zero, and in accordance with Theorem [**9.7**] any solution of either equation is a solution of the other; this is seen to be the case when we note that the second equation is the first multiplied by 2, an illustration of Theorem [**9.6**].

EXERCISES

1. Find the common solution of the equations $2x - 3y + 4 = 0$ and $x + y + 2 = 0$, using determinants, and check the result by solving the equations by the method previously known by the reader.

2. Given the equations $2x - y + 4 = 0$ and $ax - 2y + c = 0$, for what values of a and c have these equations one common solution; no common solution; an endless number of common solutions?

3. Show that

$$\begin{vmatrix} ga_1 + hd_1 & b_1 \\ ga_2 + hd_2 & b_2 \end{vmatrix} = \begin{vmatrix} ga_1 & b_1 \\ ga_2 & b_2 \end{vmatrix} + \begin{vmatrix} hd_1 & b_1 \\ hd_2 & b_2 \end{vmatrix} = g\begin{vmatrix} a_1 & b_1 \\ a_2 & b_2 \end{vmatrix} + h\begin{vmatrix} d_1 & b_1 \\ d_2 & b_2 \end{vmatrix}.$$

4. What are equations of the diagonals of the quadrilateral whose vertices are $(0, 0)$, $(a, 0)$, (a, b), $(0, b)$? Find their point of intersection and show that they bisect one another.

5. For what values of a and b are the lines $ax + 8y + 4 = 0$ and $2x + ay + b = 0$ coincident? For what values are they parallel?

6. Show that the angle ϕ between the positive directions of the perpendiculars to two intersecting lines (9.1) is given by

$$\cos \phi = \frac{a_1a_2 + b_1b_2}{e_1e_2\sqrt{(a_1^2 + b_1^2)(a_2^2 + b_2^2)}},$$

where e_1 is $+1$ or -1 so that $e_1b_1 > 0$ if $b_1 \neq 0$, and $e_1a_1 > 0$ if $b_1 = 0$; and similarly for e_2.

7. Show that the expression $-2(3x + y - 5) + 5(x - 2y - 4)$ vanishes identically when x and y take the values of the common solution of the equations $3x + y - 5 = 0$ and $x - 2y - 4 = 0$. Is it necessary to find the common solution to verify the above statement? Would the statement be equally true if the multipliers -2 and 5 were replaced by any other numbers?

8. Show that as a result of the whole discussion in § 9 we may go back and put "if and only if" in place of "if" in all the theorems of § 9 in which "if" alone appears.

9. Using Ex. 3, show that

$$\begin{vmatrix} a_1c_1 + b_1c_2 & a_1d_1 + b_1d_2 \\ a_2c_1 + b_2c_2 & a_2d_1 + b_2d_2 \end{vmatrix}$$

is equal to the product of the determinants $\begin{vmatrix} a_1 & b_1 \\ a_2 & b_2 \end{vmatrix}$ and $\begin{vmatrix} c_1 & d_1 \\ c_2 & d_2 \end{vmatrix}$.

10. The Set of Lines through a Point

From Theorem [9.2] and the other theorems in § 9 it follows that the lines with equations

$$(10.1) \qquad a_1x + b_1y + c_1 = 0, \qquad a_2x + b_2y + c_2 = 0$$

intersect, that is, meet in one point and only one point, if and only if their determinant is not equal to zero; that is, if (9.7) holds. Thus *two intersecting lines determine a point.* This is called the *dual* of the theorem that two points determine a line. Equation (5.3) gives the relation between the coordinates x, y of any point of the line determined by the points (x_1, y_1) and (x_2, y_2) and the coordinates of these two points. As the dual of this result there should be a relation between an equation of any line through a point and equations of two lines determining the point. It is this relation which we now obtain.

In connection with the equations (10.1) we consider the expression

$$(10.2) \qquad t_1(a_1x + b_1y + c_1) + t_2(a_2x + b_2y + c_2),$$

where t_1 and t_2 are constants, not both zero. Since the left-hand members of equations (10.1) are equal to zero when x and y are given the values of their common solution, it follows that the expression (10.2) is equal to zero for this common solution whatever be t_1 and t_2. Hence the common solution of equations (10.1) is a solution also of the equation

$$(10.3) \qquad t_1(a_1x + b_1y + c_1) + t_2(a_2x + b_2y + c_2) = 0.$$

This being an equation of the first degree in x and y, which is seen more clearly when it is written in the form

$$(10.4) \quad (t_1a_1 + t_2a_2)x + (t_1b_1 + t_2b_2)y + (t_1c_1 + t_2c_2) = 0,$$

we have that (10.3) for any values of t_1 and t_2, not both zero, is an equation of a line through the intersection of the lines (10.1) (see § 9, Ex. 7).

We show next that equation (10.3) for suitable values of t_1 and t_2 is an equation of any particular line through the intersection of the lines (10.1). In the first place we remark that the first of the lines (10.1) is given by (10.3) when $t_2 = 0$ and t_1 is any number different from zero, and the second of the lines (10.1) when $t_1 = 0$ and t_2 is any number different from zero. Any other line through the intersection is determined by a second point of the line, say (x_1, y_1). Equation (10.3) is an equation of this line if t_1 and t_2 are such that the expression (10.2) is equal to zero when x and y are replaced by x_1 and y_1, that is, if t_1 and t_2 are such that

$$t_1A_1 + t_2A_2 = 0, \tag{10.5}$$

where A_1 and A_2 are the numbers defined by

$$A_1 = a_1x_1 + b_1y_1 + c_1, \qquad A_2 = a_2x_1 + b_2y_1 + c_2.$$

Both of these numbers are different from zero, since (x_1, y_1) is not on either line (10.1). If we give t_1 any value other than zero and solve (10.5) for t_2, we have values of t_1 and t_2 for which (10.3) is an equation of the line of the set through (x_1, y_1). From the form of (10.5) it follows that the ratio t_1/t_2 is a fixed constant, so that, if we choose another value of t_1 and find the corresponding t_2 from (10.5), the equation (10.3) for these values differs from the equation for the former set only by a constant factor. Hence any choice of t_1 and the corresponding value of t_2 give an equation of the line, and we have the theorem

[10.1] *When the lines* (10.1) *intersect in a point, that is, when the determinant of equations* (10.1) *is not equal to zero, the equation* (10.3), *namely,*

$$t_1(a_1x + b_1y + c_1) + t_2(a_2x + b_2y + c_2) = 0,$$

for any values of the constants t_1 and t_2, not both zero, is an equation of a line through the point of intersection of the lines (10.1); *and* (10.3) *is an equation of any line through this point for suitable values of t_1 and t_2.*

Ordinarily the simplest choice of t_1 in an equation such as (10.5) is A_2, and then $t_2 = -A_1$. When these values are substituted in (10.3), we have the desired equation.

For example, if we seek an equation of the line through the point (2,1) and the intersection of the lines $x + 2y - 1 = 0$ and $3x - y + 2 = 0$, equation (10.5) is in this case $t_1 3 + t_2 7 = 0$; and consequently an equation of the line is $7(x + 2y - 1) - 3(3x - y + 2) = 0$, which reduces to $-2x + 17y - 13 = 0$.

In order to find an equation of the line through the intersection of the lines (10.1) which is parallel to the y-axis, t_1 and t_2 in (10.3) must be chosen so that the coefficient of y in (10.4) shall be equal to zero. If we take $t_1 = b_2$, $t_2 = -b_1$, and substitute these values in (10.4), the resulting equation may be written in the form

$$\begin{vmatrix} a_1 & b_1 \\ a_2 & b_2 \end{vmatrix} x + \begin{vmatrix} c_1 & b_1 \\ c_2 & b_2 \end{vmatrix} = 0, \tag{10.6}$$

which is the same as (9.4). This is not surprising, since we obtained (9.4) by multiplying the first of (9.2) by b_2 and subtracting from the resulting equation the second of (9.2) multiplied by b_1, which is equivalent to taking $t_1 = b_2$ and $t_2 = -b_1$ in (10.4). In like manner, if we take $t_1 = a_2$, $t_2 = -a_1$, the resulting equation (10.4) may be written

$$\begin{vmatrix} a_1 & b_1 \\ a_2 & b_2 \end{vmatrix} y + \begin{vmatrix} a_1 & c_1 \\ a_2 & c_2 \end{vmatrix} = 0, \tag{10.7}$$

which is the same as (9.5). Thus we have shown that the algebraic problem of finding the common solution of two equations of the first degree in x and y, when there is one, has as its geometric equivalent the finding of lines parallel to the y-axis and x-axis respectively through the point of intersection of the graphs of the two equations.

Any two lines of the set (10.3) could have been used equally well in place of the lines (10.1) as a basis for expressing an equation of every line through the point of intersection in the form (10.3), with the understanding, of course, that for a particular line the values of t_1 and t_2 depend on which two lines are used as basis. This result may be stated as follows:

[10.2] *Given all the lines through a point, each line is expressible linearly and homogeneously in terms of any two particular lines of the set.*

This is the dual of Theorem [**5.4**].

We have remarked that when $t_1 = 0$ in (10.3) the latter is equivalent to the second of equations (10.1). With this exception t_1 is not equal to zero for any of the lines (10.3), and consequently, if equation (10.3) is divided by t_1, and t_2/t_1 is replaced by t, we have as a consequence of Theorem [**10.1**] the following:

[10.3] *When the lines* (10.1) *intersect, the equation*

$$(a_1x + b_1y + c_1) + t(a_2x + b_2y + c_2) = 0, \tag{10.8}$$

where t is any constant, is an equation of a line through the intersection; and any line through the intersection, except the second line (10.1), *has an equation of the form* (10.8) *when t is given an appropriate value.*

This theorem may be proved independently by the same kind of reasoning which established Theorem [**10.1**]. In the case of Theorem [**10.3**] we have in place of (10.5) the equation $A_1 + tA_2 = 0$; and thus t is completely determined except when $A_2 = 0$, in which case we are dealing with the second of the lines (10.1). This makes the application of Theorem [**10.3**] somewhat simpler than that of Theorem [**10.1**] in some cases, and the reader may choose which theorem is better to apply in a particular problem.

From a theorem in plane geometry we have that the locus of a point equidistant from two intersecting lines is the bisectors of the angles formed by the lines. Consequently this locus consists of two of the set of lines through the point of intersection of the given lines. In order to find equations of the bisectors, we take two equations (10.1). At least one of the coefficients b_1 and b_2 is not equal to zero, otherwise the lines are parallel to one another, and to the y-axis.

We consider first the case when both b_1 and b_2 are different from zero. Let (x_1, y_1) be a point of the locus. Then its distances from the lines (10.1) are by Theorem [**8.1**]

$$\frac{a_1x_1 + b_1y_1 + c_1}{e_1\sqrt{a_1^2 + b_1^2}}, \qquad \frac{a_2x_1 + b_2y_1 + c_2}{e_2\sqrt{a_2^2 + b_2^2}}, \tag{10.9}$$

where e_1 and e_2 are $+1$ or -1 so that $e_1b_1 > 0$, $e_2b_2 > 0$. If (x_1, y_1) lies above the two lines, both the numbers (10.9) are positive and equal; if (x_1, y_1) lies below the two lines, both the numbers are negative and equal; if (x_1, y_1) lies above one of the lines and below the other, one of the numbers (10.9) is positive and the other negative, but they have the same numerical value. Since these results hold for every point on the bisectors, we have

[10.4] *The bisectors of the angles formed by two intersecting lines* (10.1) *for which* $b_1 \neq 0$ *and* $b_2 \neq 0$ *have the equations*

$$\frac{a_1x + b_1y + c_1}{e_1\sqrt{a_1^2 + b_1^2}} = \pm\frac{a_2x + b_2y + c_2}{e_2\sqrt{a_2^2 + b_2^2}}, \tag{10.10}$$

where e_1 *and* e_2 *are such that* $e_1b_1 > 0$, $e_2b_2 > 0$. *The equation with the sign* $+$ *is that of the bisector each of whose points is above, or below, both of the lines* (10.1); *and the equation with the sign* $-$ *is that of the bisector each of whose points is above one of the lines* (10.1) *and below the other.*

This theorem is illustrated in Fig. 12, where the lines (10.10) with the signs $+$ and $-$ are indicated by $(+)$ and $(-)$ respectively.

When $b_1 = 0$ or $b_2 = 0$, e_1 or e_2 must be chosen so that $e_1a_1 > 0$ or $e_2a_2 > 0$. In this case the equation (10.10) with the sign $+$ is that of the bisector whose points lie above one line and to the right of the other, or below the one and to the left of the other; and equation (10.10) with the sign $-$ is that of the bisector whose points lie above, or below, one line and to the left, or right, of the other, as the reader should verify.

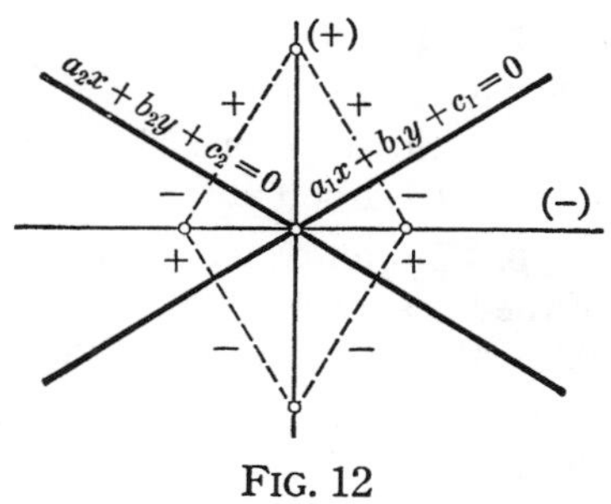

FIG. 12

For example, the bisectors of the angles between the lines

$$2x + y - 3 = 0, \quad 4x - 3y + 1 = 0$$

have the respective equations

$$(2\sqrt{5} \pm 4)x + (\sqrt{5} \mp 3)y - (3\sqrt{5} \mp 1) = 0.$$

EXERCISES

1. Find an equation of the line through the intersection of the lines $x - 2y + 7 = 0$ and $3x + y - 1 = 0$ and the point $(1, -1)$; of the line through the intersection and the point $(-2, 7)$; of the line through the intersection and parallel to the x-axis.

2. Find an equation of the line through the intersection of the lines $2x - y + 8 = 0$ and $x + y + 2 = 0$ and parallel to the line $5x - y = 0$.

3. Show by means of Theorem [**10.1**] that the line $x + 5y - 4 = 0$ passes through the intersection of the lines $2x + y + 1 = 0$ and $x - y + 2 = 0$.

4. What must be the value of a in the equation $ax + y + 6 = 0$ so that this line shall pass through the intersection of $x + y + 4 = 0$ and $2x + 3y + 10 = 0$?

5. Show that the lines $ax + by + c = 0$, whose coefficients satisfy the condition $a + b = kc$, where k is a constant not equal to zero, pass through a common point.

6. Show that when the lines (10.1) are parallel, (10.8) is an equation of a line parallel to the former for any value of t, except $t = -a_1/a_2 = -b_1/b_2$. Why does this exception not arise when the lines (10.1) intersect? Is every line parallel to the lines (10.1) so defined?

7. Find equations of the bisectors of the angles between the lines $4x - 3y + 5 = 0$ and $12x + 5y - 3 = 0$.

8. Find equations of the bisectors of the angles between the line $4x - 5y + 1 = 0$ and the line perpendicular to it through the origin.

9. Find an equation of the line through the point of intersection of the lines $5x + 2y + 1 = 0$ and $x - y - 5 = 0$ such that the point $(5, -4)$ is at the directed distance $+2$ from the line.

10. Find the point of intersection of the bisectors of the angles of the triangle whose sides have the equations

$$3x + 4y + 7 = 0, \quad 4x + 3y - 21 = 0, \quad 12x - 5y + 38 = 0,$$

that is, the center of the *inscribed* circle; find the radius of this circle.

11. Draw the graph of the lines in Ex. 10, extending the sides of the triangle, thus dividing the plane into seven compartments. Show that in three of these compartments outside the triangle there is a point which is the center of a circle tangent to the three lines; find the center and radius of each of these *escribed* circles.

11. Oblique Axes

At the beginning of § 2 we took the coordinate axes perpendicular to one another, in which case they are said to be *rectangular*, and defined the coordinates of a point by drawing through the point lines parallel to the axes. We did not define the coordinates as the perpendicular distances of the point from the y-axis and x-axis respectively, although we remarked that this is what they are when the axes are perpendicular to one another. Consequently the definition of coordinates given in § 2 applies equally well when the axes are not perpendicular to one another, in which case we say the axes are *oblique.*

The question of whether the axes are rectangular or oblique does not arise in the case of lengths parallel to the axes, and thus does not affect direction numbers, or equations of a line not involving direction cosines satisfying equation (3.6); in fact, equation (3.6) was derived from the expression (3.1) for the distance between two points (x_1, y_1) and (x_2, y_2), which presupposes that the axes are rectangular. The same assumption was involved in deriving formulas for the angle between lines and the distance from a line to a point. The reader will find it instructive, and conducive to getting a clear picture of the subject thus far, to take each theorem and determine whether its proof depends upon the axes' being rectangular.

The subject of oblique axes is introduced here not only to enable one to understand how much of what precedes is independent of the angle formed by the coordinate axes, but also because the use of such axes in certain problems leads to simpler algebraic treatment of these problems. For example, if one wishes to prove that the medians of a triangle meet in a point, it is advisable to choose two sides of the triangle as axes, and denote the vertices by $(0, 0)$, $(b, 0)$, and $(0, c)$, as shown in the accompanying figure. Equation (6.1) applies, because it expresses the equality of ratios of direction numbers.

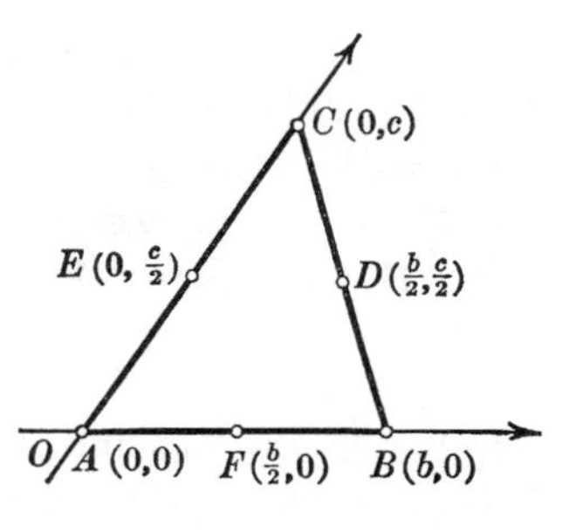

The equations of the lines BE and CF are

$$(11.1) \qquad cx + 2\,by - bc = 0, \quad 2\,cx + by - bc = 0.$$

If we subtract the first of these equations from the second, we get an equation of a line through the intersection of these lines by Theorem [**10.1**]; this equation is

$$(11.2) \qquad cx - by = 0.$$

But this is an equation of the median AD, as the reader can readily verify; and thus it is shown that the medians meet in a point. Another method of proof consists in solving (11.1) for their common solution; this is $b/3$, $c/3$, which evidently satisfies (11.2). Moreover, one shows readily that the point $(b/3, c/3)$ divides each median in the ratio 1/2.

12. The Circle

By definition a circle is the locus of points each of which is at the same distance, the radius, from a fixed point, the center. In order to find an equation of a circle, we denote by r its radius, by $P_0(x_0, y_0)$ its center, and by x, y the coordinates of a *representative* point, that is, any point on the circle. By means of Theorem [**3.1**] we obtain as an equation of the circle

$$(12.1) \qquad (x - x_0)^2 + (y - y_0)^2 = r^2,$$

which may be written

$$(12.2) \qquad x^2 + y^2 - 2\,x_0x - 2\,y_0y + (x_0^2 + y_0^2 - r^2) = 0.$$

Evidently any solution x, y of this equation (and there is an endless number of them) gives the coordinates of a point on the circle with center at P_0 and radius r. On giving x_0, y_0, r suitable values, equation (12.1) is an equation of any circle.

We remark that (12.2) is of the form

$$(12.3) \qquad x^2 + y^2 + 2\,fx + 2\,gy + k = 0.$$

Conversely, we shall determine whether an equation of this form for given values of f, g, and k is an equation of a circle. If we add $f^2 + g^2$ to the left-hand member of (12.3), the resulting expression may be written $(x + f)^2 + (y + g)^2 + k$. But then we must add $f^2 + g^2$ to the right-hand side of (12.3), so as to continue to have an equivalent equation. Consequently we can replace (12.3) by its equivalent

$$(12.4) \qquad (x + f)^2 + (y + g)^2 = f^2 + g^2 - k.$$

Comparing this equation with (12.1), we see that (12.4), and consequently (12.3), is an equation of a circle whose center is $(-f, -g)$ and whose radius is $\sqrt{f^2 + g^2 - k}$. From the form (12.4) of equation (12.3) it follows that the latter equation does not have any real solutions when $f^2 + g^2 - k < 0$. But because (12.3) is of the same form as in the case when there are real solutions, we say that when $f^2 + g^2 - k$ is negative (12.3) is an equation of an *imaginary circle*; we cannot plot such a circle, only talk about it. When $f^2 + g^2 - k = 0$, it follows from (12.4) that $x = -f$, $y = -g$ is the only real solution of (12.4), and consequently of (12.3); sometimes this point is called a *point circle,* that is, a circle of zero radius. Thus only when $f^2 + g^2 - k > 0$ is (12.3) an equation of what the reader would call a genuine circle. But if one wishes to make a general statement about the geometric significance of equation (12.3), one may say that it is an equation of a circle whatever be the values of f, g, and k, admitting the possibility of imaginary and point circles.

It is important that the reader get clearly in mind the form of equation (12.3). Note that there is no term in xy, and that the coefficients of x^2 and y^2 are equal, in this case both equal to $+1$. If they were both equal to some other number (for example 3), we could divide through by that number (which does not affect the solutions of the equation) before putting the equation in the form (12.4) by the process used above, and called completing the square of the terms involving x, and of the terms involving y.

For example, if we have the equation

$$2x^2 + 2y^2 - 5x + 4y - 7 = 0,$$

we divide by 2 and complete the squares, getting

$$x^2 - \tfrac{5}{2}x + (\tfrac{5}{4})^2 + y^2 + 2y + 1 = (\tfrac{5}{4})^2 + 1 + \tfrac{7}{2} = \tfrac{97}{16}.$$

When this equation is written in the form

$$(x - \tfrac{5}{4})^2 + (y + 1)^2 = \tfrac{97}{16},$$

it is seen that $(\frac{5}{4}, -1)$ is the center of the circle and $\frac{1}{4}\sqrt{97}$ its radius.

Geometrically we say that a line meets a circle in two points or in no points or is tangent to it. Let us see what this means algebraically. We take the equation of the line in the form

$$y = mx + h, \tag{12.5}$$

in which case m is the slope and h the y-intercept. Instead of taking a circle placed in general position with respect to the axes, we take a circle of radius r with center at the origin (for the sake of simplifying the calculations involved); its equation is

$$x^2 + y^2 = r^2. \tag{12.6}$$

The reader is familiar with the process of finding common solutions of equations such as (12.5) and (12.6), namely, substituting the expression (12.5) for y in (12.6) and solving the resulting equation. Making this substitution and collecting terms, we obtain

$$(1 + m^2)x^2 + 2\,mhx + (h^2 - r^2) = 0. \tag{12.7}$$

The geometric interpretation of this quadratic in x is that its two roots are the x-coordinates of the points in which the line meets the circle. Applying the formula for the solution of a quadratic equation to (12.7) and reducing the result, we have

$$x = \frac{-mh \pm \sqrt{r^2(1 + m^2) - h^2}}{1 + m^2}. \tag{12.8}$$

If $r^2(1 + m^2) > h^2$, (12.8) gives two real and distinct values of x. These and the corresponding values of y obtained from (12.5) on substituting these values of x are the coordinates of the two points of intersection.

If $r^2(1 + m^2) < h^2$, the two solutions (12.8) are imaginary, as are the corresponding y's. In this case the line does not meet the circle, or, as is sometimes said, it meets the circle in two imaginary points; we cannot plot such points, although we may talk about them.

We consider finally the remaining possibility, namely, $h^2 = r^2(1 + m^2)$. When $h = r\sqrt{1 + m^2}$, the two solutions of (12.7) are equal to one another, as follows from (12.8); that is, the two points coincide. In this case we say that the line is *tangent* to the circle at the (doubly counted) point P_1, whose coordinates are

$$x_1 = -\frac{mh}{1+m^2}, \quad y_1 = \frac{h}{1+m^2}, \tag{12.9}$$

as follows from (12.8) and (12.5).

The line joining the center of the circle (that is, the origin) to the point P_1 has x_1, y_1 for direction numbers, as follows from Theorem [**6.1**] (see § 6, Ex. 7); and consequently from (12.9) it follows that m, -1 are direction numbers of this line, being proportional to x_1 and y_1. In consequence of Theorem [**6.9**], m, -1 are direction numbers of a line perpendicular to the line (12.5). Hence the tangent, as defined above, is perpendicular to the radius of the circle at P_1; this agrees with the definition of a tangent to a circle, with which the reader is familiar. However, the latter definition is limited to the circle, whereas the definition we have used is general in its application.

Since similar results are obtained when h in (12.5) has the value $-r\sqrt{1+m^2}$, we have

[**12.1**] *For each value of m the two equations*

$$y = mx \pm r\sqrt{1+m^2} \tag{12.10}$$

are equations of two parallel tangents, of slope m, to the circle $x^2 + y^2 = r^2$.

This is illustrated in the accompanying figure. Equations (12.10) are equations of tangents to a circle only in case the equation of the circle has the form (12.6). However, the same process may be applied to an equation of the circle in the general form (12.3) (see Ex. 5 and § 13, Ex. 23).

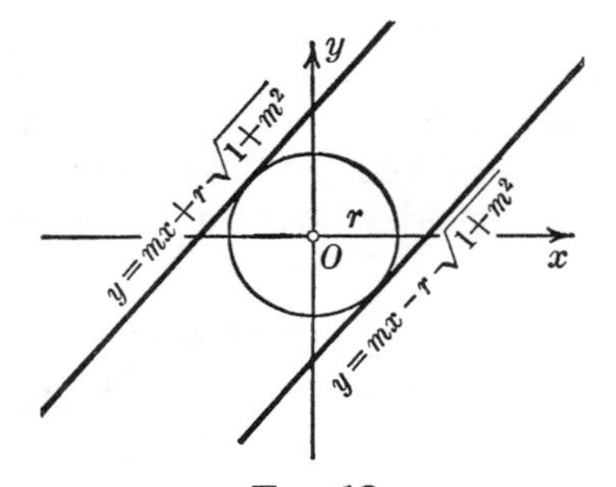

FIG. 13

When a point $P_1(x_1, y_1)$ is on the circle (12.3), the expression $x_1^2 + y_1^2 + 2fx_1 + 2gy_1 + k$ is equal to zero; but when P_1 is not on the circle, this expression is a number different from zero. We shall give a geometric interpretation of this number when P_1 lies outside the circle, as in Fig. 14. Denoting the center by C, we have from (12.4)

$$(CT)^2 = f^2 + g^2 - k.$$

Also

$$(CP_1)^2 = (x_1 + f)^2 + (y_1 + g)^2.$$

Consequently

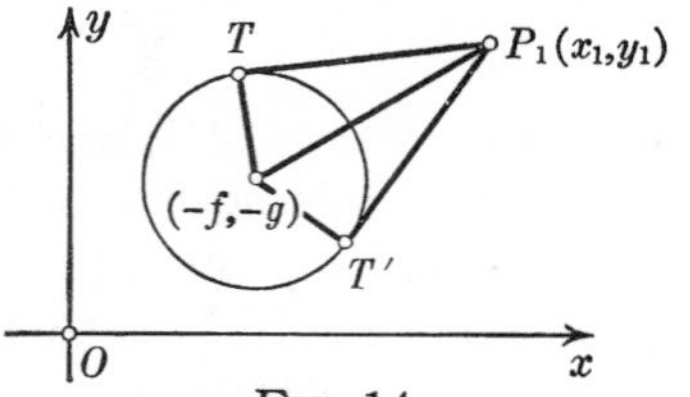

FIG. 14

$$(P_1T)^2 = (CP_1)^2 - (CT)^2 = x_1^2 + y_1^2 + 2fx_1 + 2gy_1 + k.$$

Hence, since $P_1T = P_1T'$, we have

[12.2] *When a point $P_1(x_1, y_1)$ lies outside the circle* (12.3), *the number $x_1^2 + y_1^2 + 2fx_1 + 2gy_1 + k$ is equal to the square of the distance from P_1 to the point of contact of each of the two tangents to the circle from P_1.*

Since there are three independent coefficients, f, g, k, in equation (12.3) — independent in the sense that each choice of these coefficients gives an equation of a particular circle, and a change of one or more of the coefficients chosen gives a different circle — it follows that a circle is completely determined by three conditions. The definition of a circle involves the location of its center and the length of its radius; when these are given, the values of f, g, and k are completely determined, as shown in equation (12.2). Also a circle is determined by three noncollinear points (see Ex. 8); if the coordinates of these points are given and these are substituted in (12.3), we obtain three equations to be solved for the appropriate coefficients f, g, and k in (12.3).

Consider now the two circles

$$(12.11) \qquad \begin{aligned} x^2 + y^2 + 2f_1x + 2g_1y + k_1 = 0, \\ x^2 + y^2 + 2f_2x + 2g_2y + k_2 = 0, \end{aligned}$$

and in connection with them the equation

$$(12.12) \qquad t_1(x^2 + y^2 + 2f_1x + 2g_1y + k_1) + t_2(x^2 + y^2 + 2f_2x + 2g_2y + k_2) = 0,$$

where t_1 and t_2 are constants not both zero. This equation may be written also in the form

$$(12.13) \quad (t_1 + t_2)(x^2 + y^2) + 2(t_1 f_1 + t_2 f_2)x + 2(t_1 g_1 + t_2 g_2)y + (t_1 k_1 + t_2 k_2) = 0.$$

If x_1, y_1 is a common solution of equations (12.11), it is a solution of (12.12) whatever be the values of t_1 and t_2, since the expression in each parenthesis in (12.12) becomes equal to zero when x, y are replaced by x_1, y_1. When $t_1 + t_2 \neq 0$ and equation (12.13) is divided by $(t_1 + t_2)$, the resulting equation is of the form (12.3). Consequently equation (12.12) for $t_1 + t_2 \neq 0$ is an equation of a circle passing through the points of intersection of the circles (12.11) if the circles intersect, that is, if the common solutions of (12.11) are real.

In order to find the common solutions, if any, of equations (12.11), we subtract the second of these equations from the first and obtain

$$(12.14) \quad 2(f_1 - f_2)x + 2(g_1 - g_2)y + (k_1 - k_2) = 0.$$

This does not mean that any solution of this equation is a common solution of equations (12.11), but that, if the latter have a common solution, it is a solution of (12.14). We observe that (12.14) follows from (12.13) when we take $t_2 = -t_1$ and divide out the factor t_1; consequently (12.14) is the reduced form of (12.12) when $t_2 = -t_1$. In view of this fact, if we have a common solution of (12.14) and either of equations (12.11), it is a solution of the other. In fact, if x_1, y_1 satisfy the first of (12.11) and (12.12) with $t_2 = -t_1$, then the quantity in the first parenthesis is equal to zero when x, y are replaced by x_1, y_1 throughout the equation, and consequently the expression in the second parenthesis also. Hence the problem of finding the points of intersection of the circles (12.11) reduces to that of finding the points of intersection of either circle and the line (12.14). This line is called the *radical axis* of the two circles. From considerations similar to those applied to equations (12.5) and (12.6) it follows that this line and the circles have in common two points or no real points or one point (counted doubly), in which case the two circles have the line for common tangent at the point.

When the two points of intersection of the circles (12.11) coincide, in which case the circles have a common tangent at this (doubly counted) common point, each of the circles (12.12) has the same tangent at this point.

Whether the circles have a point in common or not, the radical axis exists. From Theorem [**6.9**] and the form of (12.14) it follows that $f_1 - f_2$, $g_1 - g_2$ are direction numbers of a line perpendicular to the radical axis. But by Theorem [**6.1**] these are also direction numbers of the line joining the points $(-f_1, -g_1)$ and $(-f_2, -g_2)$, that is, the centers of the circles. Hence we have

[12.3] *Given two circles* (12.11), *equation* (12.12) *for values of* t_1 *and* t_2 *such that* $t_1 + t_2 \neq 0$ *is an equation of a circle through the points, if any, in which the circles meet; when* $t_1 + t_2 = 0$, *it is an equation of a line perpendicular to the line through the centers of the two circles.*

EXERCISES

1. Find an equation of the circle whose center is $(-2, 3)$ and whose radius is 2.

2. Find an equation of the circle of radius 3 which is tangent to the x-axis at the origin and lies below the x-axis.

3. Find the center and radius of the circle whose equation is $2x^2 + 2y^2 + 6x - 5y = 0$.

4. Show that a circle (12.3) is tangent to the x-axis, if and only if $k = f^2$. What is the condition that it be tangent to the y-axis?

5. Find equations of the tangents to the circle $x^2 + y^2 = 4$ with the slope 3; also the tangents to the circle $x^2 + y^2 + 4x - 1 = 0$ with the slope -2.

6. Find equations of the lines with the slope 3 which are at the distance 4 from the origin.

7. Find equations of the lines through the point (2, 3) which are at the distance 2 from the origin.

8. Show that the radius and center of a circle are determined by three points on the circle.

9. Prove that an angle inscribed in a semicircle is a right angle, taking $x^2 + y^2 = r^2$ as an equation of the circle.

10. Where are the points for which $x^2 + y^2 - 2x + 4y \geq 4$?

11. What is the locus with the equation

$$(2x - y + 3)(3x + 4y - 1) = 0?$$

Is the locus of $x^2 + 2xy + y^2 - 4 = 0$ a circle?

12. Find an equation of the circle through the three points (1, 1), (1, − 2), (2, 3).

13. Find the length of the tangent from the point (2, 1) to the circle $3x^2 + 3y^2 + 5x - 2y + 2 = 0$.

14. Show that an equation of any circle through the points of intersection of two circles (12.11) is given by (12.12) for suitable values of t_1 and t_2, and that the centers of all these circles lie on a line.

15. Show that an equation of any circle through the points of intersection of two circles (12.11), with the exception of the second of the circles (12.11), is given by the equation

$$x^2 + y^2 + 2f_1x + 2g_1y + k_1 + t(x^2 + y^2 + 2f_2x + 2g_2y + k_2) = 0$$

for a suitable value of t.

16. Show that equation (12.12) is an equation of a circle even if the circles (12.11) do not intersect, except when $t_1 + t_2 = 0$, and that in the latter case it is an equation of a line perpendicular to the line through the centers of the two circles.

17. Of the circles through the points of intersection of the circles $x^2 + y^2 - 4x + 2y + 4 = 0$ and $x^2 + y^2 + 6x + 8y = 0$, find

(*a*) the one which passes through the point (1, 2);
(*b*) the one whose center is on the x-axis;
(*c*) an equation of the radical axis.

18. Show that the circles (12.11) are *orthogonal* to one another, that is, that their tangents at each point of intersection are perpendicular to one another, if and only if $2(f_1f_2 + g_1g_2) = k_1 + k_2$.

19. Find an equation of the circle which passes through the point (0, 2) and is orthogonal to the two circles $x^2 + y^2 + 2x - 4y - 3 = 0$ and $x^2 + y^2 - 6x + 2y + 6 = 0$.

20. Show that, if (x_1, y_1) is a point within the circle (12.3), the expression $x_1^2 + y_1^2 + 2fx_1 + 2gy_1 + k$ is a negative number, and that its absolute value is equal to the square of one half of the chord through the point (x_1, y_1) perpendicular to the radius through this point.

21. Determine whether each of the points $(1, -\frac{1}{2})$, (2, − 1), (1, 1) is outside, on, or inside the circle $x^2 + y^2 - 3x + y + 2 = 0$.

13. Résumé. Line Coordinates

As stated in the Introduction, coordinate geometry involves the application of algebra to the study of geometric problems. The first step consists in setting up the algebraic equations (or equation) which express the conditions of a geometric problem, and solving these equations or reducing them by algebraic processes to their simplest form. The second step is the geometric interpretation of the result. Let us interpret some of the results of this chapter in the light of this statement.

A line in the plane is determined either by two of its points or by a point and the direction of the line. We have established the following equations of a line:

$$\frac{x - x_1}{x_2 - x_1} = \frac{y - y_1}{y_2 - y_1},$$

$$\frac{x - x_1}{u} = \frac{y - y_1}{v},$$

$$m(x - x_1) = y - y_1.$$

The first of these equations involves the coordinates (x_1, y_1) and (x_2, y_2) of two points on the line, the second and third a point and a direction, the latter being expressed either by direction numbers u and v, that is, numbers proportional to the cosines of the angles which the positive direction of the line makes with the positive directions of the axes, or by the slope, that is, the tangent of the angle which the positive direction of the line makes with the positive direction of the x-axis. Since $x_2 - x_1$ and $y_2 - y_1$ are direction numbers of the line, the first equation is in fact a special case of the second equation. Consequently, when the above data are given for a line, an equation of the line may be written, an equation of a line being an equation each of whose solutions consists of the coordinates of a point on the line, and conversely.

Each of the above equations is of the first degree in x and y, and it has been shown that every equation of the first degree is an equation of a line. Any such equation is of the form

$$ax + by + c = 0.$$

Suppose then that we have such an equation and wish to interpret it geometrically. If we find any solution of it, say (x_1, y_1), we can put the equation in either of the last two forms above, provided we adhere to the principle that, if in an equality of two or more ratios either term of a ratio is equal to zero, so also is the other term—a principle which has been used several times in this chapter. Also by means of a second solution the equation may be given the first of the above forms.

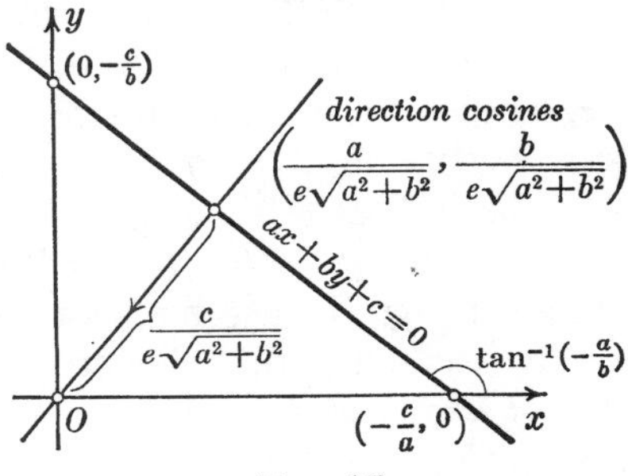

FIG. 15

Since any multiple of the above equation has the same solutions as the given equation, any such equation is an equation of the line; and consequently we should expect that not only a, b, and c but also quantities proportional to them have geometric significance. To overcome this ambiguity we write the above equation in one of the following forms, of which the first is possible only when a, b, and c are all different from zero, and the second when $b \neq 0$:

$$\text{(13.1)} \qquad \frac{x}{-\frac{c}{a}} + \frac{y}{-\frac{c}{b}} = 1; \text{ in other notation, } \frac{x}{g} + \frac{y}{h} = 1.$$

$$\text{(13.2)} \qquad y = -\frac{a}{b}x - \frac{c}{b}; \text{ in other notation, } y = mx + h.$$

$$\text{(13.3)} \qquad \frac{a}{e\sqrt{a^2+b^2}}x + \frac{b}{e\sqrt{a^2+b^2}}y + \frac{c}{e\sqrt{a^2+b^2}} = 0,$$

where e is $+1$ or -1 so that $eb > 0$ when $b \neq 0$, and $ea > 0$ when $b = 0$.

In (13.1) g and h are the x- and y-intercepts respectively; in (13.2) m is the slope of the line and h its y-intercept; in (13.3) $a/e\sqrt{a^2+b^2}$ and $b/e\sqrt{a^2+b^2}$ are the direction cosines of any of the endless number of lines perpendicular to the given line, and $c/e\sqrt{a^2+b^2}$ is the directed distance of the origin from the line (see Fig. 15).

Since the coefficients a, b, and c of an equation $ax + by + c = 0$ determine the equation, and consequently the line of which it is an equation, we call a, b, and c *line coordinates* of the line. Since an equation of a line is determined only to within a constant multiplier, it follows that if a, b, and c are line coordinates of a line, so also are ka, kb, and kc line coordinates of the same line for every value of k different from zero. The line coordinates $1/g$, $1/h$, -1 in (13.1), m, -1, h in (13.2), and $a/e\sqrt{a^2 + b^2}$, $b/e\sqrt{a^2 + b^2}$, and $c/e\sqrt{a^2 + b^2}$ in (13.3), have in each case the geometric significance mentioned in the preceding paragraph.

Now we may state Theorem [**10.1**] as follows: *Line coordinates of any line through the intersection of two lines with coordinates* a_1, b_1, c_1 *and* a_2, b_2, c_2 *are of the form* $t_1a_1 + t_2a_2$, $t_1b_1 + t_2b_2$, $t_1c_1 + t_2c_2$. Thus line coordinates of each of the set of all the lines through a point are expressible linearly and homogeneously in terms of line coordinates of any two lines of the set. This is the dual of Theorem [**5.4**].

If the reader at some time takes up the study of projective geometry, he will find that in this subject line coordinates are as fundamental as point coordinates, and that corresponding to any theorem concerning points and lines there is a dual theorem concerning lines and points.

The two steps involved in coordinate geometry mentioned in the first paragraph of this section are involved in the solution of any locus problem, that is, in finding the answer to the question: What is the locus of a point satisfying certain geometric conditions? First of all the reader is expected to set up the equation (or equations) satisfied by the coordinates of any and every point meeting the conditions of the problem. Ordinarily these conditions involve the distance between points, or the distance of a point from a line, or the angle between two lines, formulas for which have been derived in the text. The next step, after the equations have been solved or reduced to simple forms, is to interpret the result geometrically, stating whether the locus is a line, a circle, or some other curve, as the reader can do when he knows the forms of equations of other curves, knowledge which he will acquire in later parts of this book.

As an example of the foregoing we refer to equations (10.10), obtained as answer to the question: What is the locus of a point equidistant from the intersecting lines (10.1)? We observed that it consists of two lines, the bisectors of the angles between the two given lines. Suppose again that we were asked to find the locus of a point satisfying certain geometric conditions and eventually obtained an equation of the form

$$x^2 + y^2 + 2fx + 2gy + k = 0.$$

We should know from § 12 that the locus is a circle and how to specify its center and radius.

The résumé given in the first part of this section is in no sense a complete summary of this chapter. A good way for the reader to get a picture of the chapter as a whole is by reading successively the theorems given; the same suggestion applies to succeeding chapters.

In this chapter use has been made of theorems of plane geometry assumed to be known by the reader, and the results of their use have been translated into algebraic form. There has been no attempt to develop coordinate geometry systematically from a set of axioms. However, in the Appendix to Chapter I (p. 279) the reader will find an exposition of the relation between a set of axioms for Euclidean plane geometry and coordinate geometry as developed in the present chapter.

EXERCISES

1. Find the locus of a point such that 3 times its distance from the y-axis plus 4 times its distance from the x-axis is equal to 12.

2. Find the locus of a point whose distance from the line $y - b = 0$ is a constant times its distance from the y-axis.

3. Find the locus of a point the ratio of whose distances from lines through the point (x_1, y_1) parallel to the y-axis and x-axis respectively is equal to u/v, where u and v are constants.

4. Given the isosceles right-angled triangle whose vertices are $(0, 0)$, $(a, 0)$, and $(0, a)$, find the locus of a point the square of whose distance from the hypotenuse is equal to the product of its distances from the two legs of the triangle.

5. Find the locus of the mid-point of a line of constant length having its extremities on the positive parts of the axes.

6. Find the locus of a point which is twice as far from the line $2x - 3y + 1 = 0$ as from the line $x - 2y + 4 = 0$.

7. Find the locus of a point such that, if Q and R are its projections on the coordinate axes, the distance QR is constant.

8. Find the locus of a point such that the mid-point of the segment joining it to the point $(0, b)$ lies on the circle $x^2 + y^2 = a^2$.

9. Show that the locus of a point whose distances from two fixed points $P_1(x_1, y_1)$ and $P_2(x_2, y_2)$ are in constant ratio different from unity is a circle. When this ratio is equal to unity, what is the locus?

10. Find the locus of a point the sum of whose distances from the sides of the triangle $y = 0$, $3y - 4x = 0$, $12x + 5y - 60 = 0$ is constant.

11. What is the locus of a point the sum of whose distances from any number n of lines is constant?

12. What is the locus of a point the sum of the squares of whose distances from any number n of points is constant?

13. Find two points on the y-axis four units distant from the line $3x + 4y - 12 = 0$.

14. The base of an equilateral triangle lies on the line $3x - 2y + 5 = 0$ and the opposite vertex is $(4, -1)$. Find equations of the other two sides of the triangle.

15. Show that the four lines

$$ax + by + c = 0, \qquad -bx + ay + c = 0,$$
$$-ax - by + c = 0, \qquad bx - ay + c = 0$$

form a square with center at the origin. What is the situation when the axes are oblique?

16. Find the distance of the center of the circle

$$x^2 + y^2 - 2x + 4y - 4 = 0$$

from the line $4x - 3y + 6 = 0$, and determine thereby whether this line lies above or below the circle, intersects the circle, or is tangent to it.

17. Show that if a point P lies above or below each of two intersecting lines, the angle at P between the positive directions of the perpendiculars upon the lines is the supplement of the angle in which

P lies; that if P lies above one of the lines and below the other, the angle between the positive directions of the perpendiculars is equal to the angle in which P lies.

18. Find what relation must hold in each case between the coefficients of the equation $ax + by + c = 0$ of a line in order that the line

(*a*) have the intercept 2 on the x-axis;
(*b*) have equal intercepts;
(*c*) be perpendicular to the line $2x - 3y + 1 = 0$;
(*d*) pass through the origin;
(*e*) pass through the point $(5, -4)$;
(*f*) be at the directed distance -3 from the origin;
(*g*) be at the directed distance 4 from the point $(2, 3)$;
(*h*) pass through the intersection of the lines $3x + y - 2 = 0$ and $x - 2y + 1 = 0$.

19. Find equations of the lines through the point $(2, 3)$ which form with the axes a triangle of area 16.

20. Given the triangle whose vertices are $O(0, 0)$, $A(a, 0)$, $B(0, b)$, prove that the line joining the mid-points of any two sides is parallel to the third side and equal to one half of it; that, if on the line joining O to the mid-point C of AB we take any point P and denote by D and E the points in which AP meets OB and BP meets OA respectively, then DE is parallel to AB. Which, if any, of the foregoing results hold when the axes are oblique?

21. Given the triangle whose vertices are $A(a, 0)$, $B(0, b)$, and $C(c, 0)$, prove that the perpendiculars from the vertices of the triangle upon the opposite sides meet in a point; that the perpendicular bisectors of the sides meet in a point.

22. Given the rectangle whose vertices are $O(0, 0)$, $A(a, 0)$, $C(a, b)$, $B(0, b)$, if E and F are mid-points of OB and AC, prove that the lines AE and BF trisect the diagonal OC. Does this result hold for the figure $OACB$ when the axes are oblique?

23. Find the condition upon the coefficients in the equations (12.3) of a circle and (12.5) of a line so that the line shall be tangent to the circle.

24. Prove that the radical axes of three circles whose centers are not all on the same line meet in a point.

25. Find an equation of the circle circumscribing the triangle whose sides are $x = 0$, $\quad y = 2x$, $\quad 2x + y - 8 = 0$.

26. Find an expression for the area of a triangle whose vertices are (x_1, y_1), (x_2, y_2), and (x_3, y_3).

27. Prove that the line $x + \sqrt{3}\,y = 0$ bisects the triangle whose vertices are $(-3\sqrt{3}, 3)$, $(-1, -\sqrt{3})$, and $(1, \sqrt{3})$.

28. Find an equation of the circle which passes through the point $(\frac{3}{5}, -\frac{4}{5})$, is tangent to the line $y = \sqrt{3}\,x + 2$, and whose center is on the line $x - 2y = 0$.

29. Find the coordinates of the point equidistant from the lines $5x - 12y - 5 = 0$ and $4x - 3y - 4 = 0$, and at the distance 2 from the origin.

30. Show that the curve with the equation

$$(2x - y + 2)(3x + 4y - 1) + (2x - y + 2)(x + y) + (3x + 4y - 1)(x + y) = 0$$

passes through the vertices of the triangle whose sides have the equations

$$2x - y + 2 = 0, \quad 3x + 4y - 1 = 0, \quad x + y = 0.$$

31. What is the significance of equations (10.10) when the lines (10.1) are parallel?

"

CHAPTER 2

Lines and Planes in Space. Determinants

"

II

14. Rectangular Coordinates in Space

The reader will agree that the position of any point in a room is fixed by its distances from the floor and from two adjacent walls. If the planes of the floor and the two walls are thought of as extended indefinitely, the position of any point in space is fixed by its distances from these three planes, provided that one has a means of indicating whether a point is on one side or the other of a plane. A similar question arose in the case of the points in the plane and was met in § 2 by using the concept of positive and negative distances from two perpendicular lines, the coordinate axes. We generalize this concept to points in space, and take as the basis of a coordinate system three planes meeting in a point O, every two planes being perpendicular to one another, just as in the case of the floor and two adjacent walls of a room. Fig. 16 is the customary way of drawing three mutually perpendicular planes, the line Oy pointing in the general direction of the observer. In Fig. 16 the lines of intersection of the planes, namely, Ox, Oy, and Oz, are called the *x-axis*, *y-axis*, and *z-axis* respectively. The plane of the two lines Ox and Oy is called the *xy-plane*; similarly we have the *yz-plane* and the *xz-plane*. These are called the *coordinate planes*.

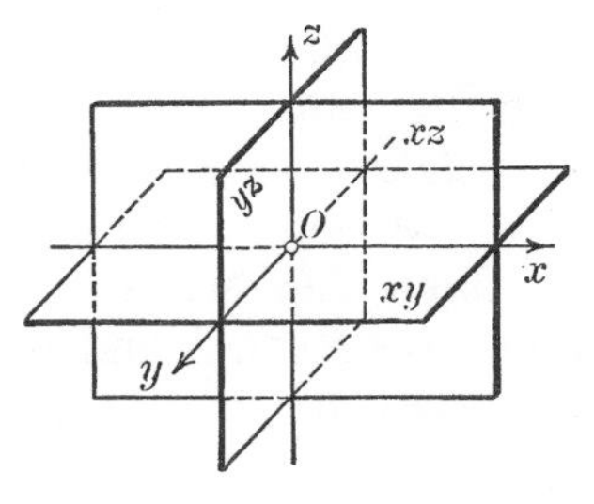

FIG. 16

The figure corresponds to the case in which the eye of the reader is in front of the xz-plane, to the right of the yz-plane, and above the xy-plane. Distances measured upward from the xy-plane are taken as positive, downward as negative; to the right of the yz-plane as positive, to the left as negative; in front of the xz-plane as positive, and back of it as negative. The reader will get a picture of this situation by looking at a lower corner of the room, and calling it O; the floor lines through O to the right and left correspond to the x-axis and y-axis respectively, and the line of intersection of the two walls through O to the z-axis. The three planes divide space into eight parts, called *octants*, just as eight rooms meet in one point when the room in which the reader is sitting is on the second floor and the two adjacent walls at O are inside walls.

Coordinate planes and coordinate axes having been chosen as just described, the coordinates of a given point P are defined as follows: Draw through P three planes parallel to the coordinate planes, and denote by P_x, P_y, and P_z the points in which these planes meet the x-axis, y-axis, and z-axis respectively. A unit of length having been chosen, the lengths of the directed segments OP_x, OP_y, and OP_z are by definition the x-, y-, and *z-coordinates* of P. Each of these coordinates is positive or negative according as the directed line segment on the corresponding coordinate axis is positive or negative, as defined in the preceding paragraph.

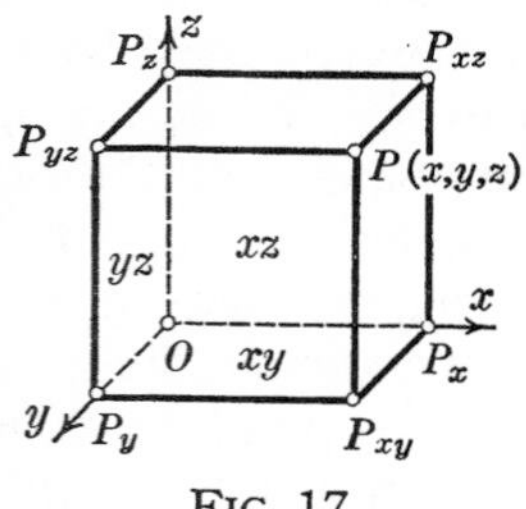

FIG. 17

Conversely, given any three numbers x, y, z, to find the point of which they are coordinates we lay off from O on the coordinate axes Ox, Oy, and Oz distances of x, y, and z units respectively, in the positive or negative direction in each case according as the respective numbers are positive or negative, and denote the end points by P_x, P_y, and P_z respectively. Through these points we draw planes parallel to the yz-, xz-, and xy-planes respectively; the point P of intersection of these planes is at the directed distances x, y, z from the coordinate planes, since parallel planes are everywhere equally distant. Consequently P is the point with the coordinates x, y, z and is indicated by $P(x, y, z)$.

An equivalent way of locating $P(x, y, z)$ is as follows: Lay off a length of x units on the x-axis in the appropriate direction from O; at the end point P_x of this segment draw a line in the xy-plane perpendicular to the x-axis; and on this line lay off from P_x in the appropriate direction a length of y units. At the end point P_{xy} of this segment construct a line parallel to the z-axis, that is, perpendicular to the xy-plane, and on this line lay off from P_{xy} in the appropriate direction a length of z units. The end of this segment is the point $P(x, y, z)$.

The coordinate planes, as defined, being mutually perpendicular, the x-, y-, and z-coordinates of a point are the perpendicular distances of the point from the yz-, xz-, and xy-planes respectively. There may be times when it is advantageous to take as coordinate planes three intersecting planes not mutually perpendicular; in such cases the coordinates as defined in the preceding paragraph are not the perpendicular distances of the point from the planes (see § 11). Throughout this book mutually perpendicular planes are used as coordinate planes.

The point in which the perpendicular from a point P upon a plane meets the plane is called the *orthogonal projection of P upon the plane*; thus, P_{xy} in Fig. 17 is the orthogonal projection of P upon the xy-plane. The point in which the perpendicular from a point P upon a line meets the line is called the *orthogonal projection of P upon the line*; thus, P_x is the orthogonal projection upon the x-axis of the points P, P_{xy}, and P_{xz}. The *projection of a line segment upon a line*, or *plane*, is the line segment whose end points are the orthogonal projections of the end points of the given segment upon the line, or plane.

Two points P_1 and P_2 are said to be *symmetric with respect to a point* when the latter bisects the segment P_1P_2; *symmetric with respect to a line* when the latter is perpendicular to the segment P_1P_2 and bisects it; *symmetric with respect to a plane* when the latter is perpendicular to the segment P_1P_2 and bisects it. The reader can picture each of these situations geometrically, but he is not in position yet to handle any one of these types of symmetry algebraically for general positions of the points P_1 and P_2. However, he is in position to discuss symmetry with respect to the origin, the coordinate axes, and the coordinate planes (see Ex. 3).

It is evident that it is not possible to represent a spatial figure accurately on a plane. However, it is convenient to use what is called *parallel projection*. In applying this method, figures in the xz-plane and in planes parallel to it are represented as they are; for example, the point $(x_1, 0, z_1)$ in the xz-plane is placed at the distances x_1 and z_1 from the z-axis and x-axis respectively. On the other hand, the y-axis and lines parallel to it are drawn to make the angle 135° with the x-axis and z-axis as in Figs. 16 and 17, and lengths in this direction are foreshortened by the factor $1/\sqrt{2}$; that is, for a length l in this direction the length $l/\sqrt{2}$ is laid off on the y-axis or a line parallel to it, as the case may be. Thus, if squared coordinate paper is used, distances along or parallel to the x-axis or z-axis are laid off to their full amount with the side of a square as unit, whereas distances along or parallel to the y-axis are laid off along a diagonal of a small square. On the paper the length of a diagonal is $\sqrt{2}$, but since $1/\sqrt{2}$ is the unit of length along a diagonal in accordance with the described method of representation, in such representation the length of the diagonal is 2; for, $\sqrt{2} \Big/ \frac{1}{\sqrt{2}} = 2$. The angle 135° and the resulting factor $1/\sqrt{2}$ are chosen because they are simple to handle, and because they have been found to give a clear conception of a spatial figure.

This chapter deals with configurations of points, lines, and planes in space defined geometrically, these definitions being then translated into algebraic form involving the coordinates of a representative point of the locus under consideration. In the definition of coordinates we have used the concepts of parallel and perpendicular planes, and of a line as the intersection of two planes, and have asked the reader to visualize the definition by considering the walls and floor of a room as planes, and their intersections as lines. In Chapter 1 we defined a line as a locus determined by any two of its points and having the same direction throughout. Also we explained what was meant by saying that a plane is a two-dimensional locus, but we did not define a plane. When we look upon a plane as lying in space, we use Euclid's definition that a plane is a two-dimensional locus such that every point of a line which has two points in the locus is in the locus. A sphere is two-dimensional in the sense that any point on it can be fixed by two numbers, for example, by latitude and longitude; but a sphere does not possess the above property, nor does any surface other than a plane.

In the consideration of points, lines, and planes in general position with respect to the coordinate axes, we are concerned with such metric questions as the distance between two points, the angle between two lines or two planes, and the distance of a point from a plane. These quantities having been expressed in algebraic form, the reader is enabled to convert a geometric problem into an algebraic one, and by algebraic processes, frequently by the use of determinants, to arrive at the solution of a problem more readily than had he employed purely geometric reasoning.

EXERCISES

1. What are the coordinates of the points O, P_x, P_y, P_z, P_{xy}, P_{yz}, and P_{xz} in Fig. 17?

2. Show that in Fig. 17 the line segment OP_x is the projection of the segment OP on the x-axis, and OP_{xz} the projection of OP on the xz-plane.

3. What are the coordinates of the seven points whose coordinates have the same numerical values as those of the point (1, 2, 3) but with one or more of the coordinates negative? Which pairs of these eight points are symmetric with respect to the origin; with respect to the y-axis; with respect to the xz-plane?

4. Where are the points for which $x = y$; for which $x < y$; for which $x = 3$ and $y = 2$?

5. Where are the points for which $x^2 + y^2 = 4$; for which $x^2 + y^2 + z^2 > 9$?

6. Given the points $P_1(1, -3, 4)$, $P_2(4, 2, -2)$, $P_3(0, 1, 5)$, and $P_4(6, 5, -3)$, show that the sum of the projections of the line segments P_1P_2, P_2P_3, and P_3P_4 upon the x-axis is equal to the projection of P_1P_4 on this axis, and that the same is true of the projections on the y-axis and the z-axis. Is this result true for any four points whatever? Is it true for any number of points?

7. A cube of side 5 has one vertex at the point (1, 0, 0) and the three edges from this vertex respectively parallel to the positive x-axis, and the negative y- and z-axes. Find the coordinates of the other vertices and of the center of the cube.

8. Plot to scale on a single sheet of squared paper the following nine points: (3, 2, 3), (3, 4, 0), (3, 0, 0), (3, 0, 3), (3, −4, 3), (3, −2, 0), (3, −2, −1), (3, 0, −1), (3, 3, −1).

9. Find an equation of the locus of a point which is twice as far from the xy-plane as from the xz-plane. From geometric considerations what is this locus?

15. Distance between Two Points. Direction Numbers and Direction Cosines of a Line Segment. Angle between Two Line Segments

In deriving the formula for the distance between two points $P_1(x_1, y_1, z_1)$ and $P_2(x_2, y_2, z_2)$ we make use of Fig. 18. The reader may get a good idea of this figure if he holds a box in the room with its edges parallel to the three lines of intersection of two adjacent walls and the floor, considered to be the coordinate planes. We note that the angle P_1QP_2 is a right angle, and consequently

$$(P_1P_2)^2 = (P_1Q)^2 + (QP_2)^2.$$

Also, the angle QRP_2 being a right angle, we have

$$(QP_2)^2 = (QR)^2 + (RP_2)^2.$$

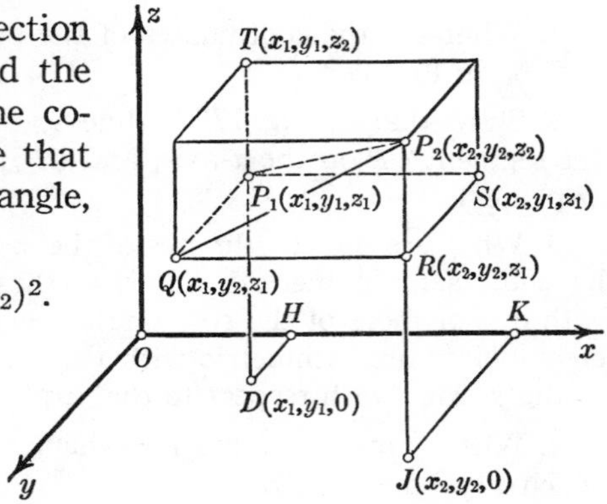

FIG. 18

Since

$$(15.1) \qquad QR = x_2 - x_1, \qquad P_1Q = y_2 - y_1, \qquad RP_2 = z_2 - z_1,$$

we have the following expression for the square of the distance P_1P_2:

$$(15.2) \qquad \begin{aligned}(P_1P_2)^2 &= (QR)^2 + (P_1Q)^2 + (RP_2)^2 \\ &= (x_2 - x_1)^2 + (y_2 - y_1)^2 + (z_2 - z_1)^2.\end{aligned}$$

If the segment P_1P_2 were placed in any other octant, some of the coordinates of P_1 and P_2 would be negative, and also some of the expressions in the above parentheses might be negative, but the above formula would still hold true. When the line is parallel to the xy-plane, $z_2 - z_1 = 0$, and the above formula reduces to the one in § 3, as we should expect. When the line is parallel to either of the other coordinate planes, the corresponding term in (15.2) is equal to zero. Hence we have the following theorem:

[15.1] *The distance between the points* (x_1, y_1, z_1) *and* (x_2, y_2, z_2) *is*

$$\sqrt{(x_2 - x_1)^2 + (y_2 - y_1)^2 + (z_2 - z_1)^2}. \tag{15.3}$$

This is a generalization of Theorem [3.1] in the plane.

The length and sign of the directed segments QR, P_1Q, and RP_2 are given by (15.1) no matter in which octant P_1 lies and in which P_2 lies. These numbers determine the "box" of which P_1P_2 is a diagonal, and consequently determine the direction of P_1P_2 relative to the coordinate axes. They are called *direction numbers* of the line segment. In like manner $x_1 - x_2$, $y_1 - y_2$, and $z_1 - z_2$ are direction numbers of the line segment P_2P_1. Thus a line segment has two sets of direction numbers, each associated with a *sense along the segment* and either determining the direction of the segment relative to the coordinate axes. But a *sensed* line segment, that is, a segment with an assigned sense, has a single set of direction numbers.

Any other line segment parallel to P_1P_2 and having the same length and sense as P_1P_2 has the same direction numbers as P_1P_2; for, this new segment determines a "box" equal in every respect to the one for P_1P_2. This means that the differences of the x's, y's, and z's of the end points of such a parallel segment are equal to the corresponding differences for P_1 and P_2. Since one and only one line segment having given direction numbers can be drawn from a given point, we have that a sensed line segment is completely determined by specifying its initial point and its direction numbers.

There is another set of numbers determining the direction of a line segment, which are called the *direction cosines* of the line segment, whose definition involves a convention as to the positive sense along the segment. This convention is that, when a segment is not parallel to the xy-plane, upward along the segment is the *positive sense* on the segment; when a segment is parallel to the xy-plane but not parallel to the x-axis, toward the observer in Fig. 18 is the *positive sense* on the segment (this is the convention used in § 3); when the segment is parallel to the x-axis, toward the right is the *positive sense*. In Fig. 18 the distance P_1P_2 is a positive number, being measured in the positive sense, and the distance P_2P_1 is a negative number (the absolute values of these two numbers are the same), just as distances measured on the x-axis to the right, or left, of a point on the axis are positive, or negative. When the sense on a line segment is assigned by this convention, we refer to it as a *directed* line segment.

By definition *the direction cosines of a line segment are the cosines of the angles of the positive direction of the segment and the positive directions of the x-, y-, and z-axes respectively; that is, of the angles made by the positive direction of the segment with line segments drawn through any point of the given segment parallel to and in the positive directions of the coordinate axes.* Thus in Fig. 18, if we denote by A, B, and C the angles SP_1P_2, QP_1P_2, and TP_1P_2, the direction cosines λ, μ, and ν (nu) of the line segment P_1P_2 are given by

$$\lambda = \cos A, \quad \mu = \cos B, \quad \nu = \cos C. \tag{15.4}$$

If we denote the positive distance P_1P_2 by d, we have from (15.1) and (15.4)

$$x_2 - x_1 = d\lambda, \quad y_2 - y_1 = d\mu, \quad z_2 - z_1 = d\nu, \tag{15.5}$$

since $QR = P_1S = P_1P_2 \cos A$, and so on. If now we imagine P_1 and P_2 interchanged in Fig. 18, this does not alter λ, μ, and ν, since their values depend only upon the direction of the segment relative to the coordinate axes; consequently in this case equations (15.5) become

$$x_1 - x_2 = \bar{d}\lambda, \quad y_1 - y_2 = \bar{d}\mu, \quad z_1 - z_2 = \bar{d}\nu,$$

where $\bar{d}$ is the distance P_2P_1 in the new figure and is positive. Hence equations (15.5) hold also if d is negative, that is, when P_1 is above P_2 on the line, in which case the distance P_1P_2 is negative.

We consider now the difference, if any, in the above results when the positive sense of the segment makes an obtuse angle with the x-axis or the y-axis. If the angle A for the segment P_1P_2 is obtuse, then $x_2 < x_1$ and the first of equations (15.5) is satisfied, since $\cos A$ is negative in this case; similarly when B is obtuse. Consequently equations (15.5) hold in every case, with the understanding that d is the sensed distance P_1P_2.

Thus a line segment is completely determined by an end point, its direction cosines, and its directed length measured from the given end point. Parallel line segments have the same direction cosines, and direction numbers of two such segments are equal or proportional, as follows from (15.5).

From the definition of the positive sense of a line segment it follows that when P_1P_2 is not parallel to the xy-plane ν is positive, whereas λ and μ may take any values between -1 and $+1$; when P_1P_2 is parallel to the xy-plane but not to the x-axis, $\nu = 0$, μ is positive, and λ may take any value between -1 and $+1$; when P_1P_2 is parallel to the x-axis, $\nu = \mu = 0$, $\lambda = 1$.

When the expressions for $x_2 - x_1$, and so on, from (15.5) are substituted in (15.2), and we note that $(P_1P_2)^2$ is d^2, we obtain

$$\lambda^2 + \mu^2 + \nu^2 = 1. \tag{15.6}$$

The converse of this result may be established as in the case of Theorem [**3.2**]. Accordingly we have the theorem

[**15.2**] *The direction cosines λ, μ, ν of any line segment satisfy the equation $\lambda^2 + \mu^2 + \nu^2 = 1$; ν is never negative; when $\nu > 0$, $-1 < \lambda < 1$, $-1 < \mu < 1$; when $\nu = 0$ and $\mu \neq 0$, then $\mu > 0$, $-1 < \lambda < 1$; when $\nu = \mu = 0$, $\lambda = 1$; and, conversely, any numbers λ, μ, ν satisfying these conditions are direction cosines of a line segment.*

Consider in connection with the line segment P_1P_2 another line segment P_1P_3, where P_3 is the point (x_3, y_3, z_3). We denote by θ the angle formed at P_1 by the sensed segments P_1P_2 and P_1P_3; and we note that if they have the same direction, $\theta = 0°$ or $180°$ according as the segments have the same or opposite sense. We consider the case when the two segments do not have the same direction, and denote by λ_1, μ_1, ν_1 and λ_2, μ_2, ν_2 the direction cosines of the segments P_1P_2 and P_1P_3 respectively, and by d_1 and d_2 their respective directed lengths. If we draw from the origin two line segments OP_1' and OP_2' parallel to and of the same directed lengths as P_1P_2 and P_1P_3 respectively, we have a graph like Fig. 4, with the difference that the points P_1' and P_2' are $(d_1\lambda_1, d_1\mu_1, d_1\nu_1)$ and

$(d_2\lambda_2, d_2\mu_2, d_2\nu_2)$. Now if l_1 and l_2 again denote the numerical lengths of OP_1' and OP_2', we have equation (3.7), but in place of (3.8) we have

$$\begin{aligned} l^2 &= (d_2\lambda_2 - d_1\lambda_1)^2 + (d_2\mu_2 - d_1\mu_1)^2 + (d_2\nu_2 - d_1\nu_1)^2 \\ &= d_2^2(\lambda_2^2 + \mu_2^2 + \nu_2^2) + d_1^2(\lambda_1^2 + \mu_1^2 + \nu_1^2) \\ &\qquad - 2\, d_1 d_2(\lambda_1\lambda_2 + \mu_1\mu_2 + \nu_1\nu_2) \\ &= l_2^2 + l_1^2 - 2\, d_1 d_2(\lambda_1\lambda_2 + \mu_1\mu_2 + \nu_1\nu_2), \end{aligned}$$

the last expression being a consequence of Theorem [**15.2**] and the fact that $l_1^2 = d_1^2$ and $l_2^2 = d_2^2$. From (3.7) and the above expression for l^2 we have

$$l_1 l_2 \cos\theta = d_1 d_2(\lambda_1\lambda_2 + \mu_1\mu_2 + \nu_1\nu_2). \tag{15.7}$$

If d_1 and d_2 are both positive or both negative, $l_1 l_2 = d_1 d_2$; if d_1 and d_2 differ in sign, $l_1 l_2 = - d_1 d_2$. Hence we have

[**15.3**] *The angle θ between two directed line segments which have one end point in common and whose respective direction cosines are λ_1, μ_1, ν_1 and λ_2, μ_2, ν_2 is given by*

$$\cos\theta = e(\lambda_1\lambda_2 + \mu_1\mu_2 + \nu_1\nu_2), \tag{15.8}$$

where e is $+1$ or -1 according as the two segments have the same sense (both positive or both negative) or opposite sense.

From equations (15.7) and (15.5) we have

[**15.4**] *The angle θ between the line segments from the point (x_1, y_1, z_1) to the points (x_2, y_2, z_2) and (x_3, y_3, z_3) is given by*

(15.9)

$$\cos\theta = \frac{(x_2 - x_1)(x_3 - x_1) + (y_2 - y_1)(y_3 - y_1) + (z_2 - z_1)(z_3 - z_1)}{l_1 l_2},$$

where l_1 and l_2 are the respective lengths (not directed distances) of the segments P_1P_2 and P_1P_3.

From (15.8) and (15.6) we have

$$\begin{aligned} \sin^2\theta = 1 - \cos^2\theta &= (\lambda_1^2 + \mu_1^2 + \nu_1^2)(\lambda_2^2 + \mu_2^2 + \nu_2^2) \\ &\qquad - (\lambda_1\lambda_2 + \mu_1\mu_2 + \nu_1\nu_2)^2 \\ &= \begin{vmatrix} \mu_1 & \nu_1 \\ \mu_2 & \nu_2 \end{vmatrix}^2 + \begin{vmatrix} \nu_1 & \lambda_1 \\ \nu_2 & \lambda_2 \end{vmatrix}^2 + \begin{vmatrix} \lambda_1 & \mu_1 \\ \lambda_2 & \mu_2 \end{vmatrix}^2, \end{aligned} \tag{15.10}$$

as may be verified by multiplying out and reducing the first of these expressions for $\sin^2\theta$ and expanding the second.

EXERCISES

1. Find the vertices of the "box" of which the line segment from $P_1(1, -2, 3)$ to $P_2(2, 4, -1)$ is a diagonal; find also the direction numbers and direction cosines of this segment, and the directed distance P_1P_2.

2. Find a so that the line segments from the point $(-3, 1, 5)$ to the points $(a, -2, 3)$ and $(1, 2, -4)$ make an angle whose cosine is $4/5$; so that the two segments shall be perpendicular.

3. Find the condition to be satisfied by the coordinates of the points (x_1, y_1, z_1) and (x_2, y_2, z_2) in order that the line segment with these as end points shall subtend a right angle at the origin.

4. Show that the points (x_1, y_1, z_1) and (x_2, y_2, z_2) are collinear with the origin, if and only if their coordinates are proportional.

5. Let $P(x, y, z)$ be the point on the line segment P_1P_2 with end points (x_1, y_1, z_1) and (x_2, y_2, z_2) which divides the segment in the ratio h_1/h_2; denote by Q_1, Q, Q_2 the orthogonal projections of P_1, P, P_2 on the xy-plane; show that Q divides the line segment Q_1Q_2 in the ratio h_1/h_2, and that similar results hold when P_1, P, P_2 are projected orthogonally on the yz- and xz-planes. Derive from these results and equations (4.5) and (4.6) the first two of the following expressions for the coordinates of P in terms of the coordinates of P_1 and P_2, and h_1 and h_2:

$$\text{(i)} \qquad x = \frac{h_2x_1 + h_1x_2}{h_1 + h_2}, \qquad y = \frac{h_2y_1 + h_1y_2}{h_1 + h_2}, \qquad z = \frac{h_2z_1 + h_1z_2}{h_1 + h_2},$$

and derive the third by projection upon the yz- and xz-planes.

6. Show that, when h_1 or h_2 in equations (i) of Ex. 5 is a negative number, these equations give the coordinates of a point on an extension of the line segment beyond P_1 or beyond P_2 according as $|h_1|$ is less or greater than $|h_2|$, and that P_1P and PP_2 are in the ratio h_1/h_2; in this case P is said to divide the segment P_1P_2 *externally*, and when h_1/h_2 is positive, *internally*.

7. Find the coordinates of the mid-point of the line segment P_1P_2 of Ex. 5, and in particular of the line segment joining the points $(3, -2, 3)$ and $(2, 2, -3)$. Find the coordinates of the points where a segment is trisected.

8. In what ratio is the line segment joining the points (2, 3, 1) and (1, 5, − 2) cut by its intersection with each of the *xy*-, *yz*-, and *xz*-planes respectively, and what are the coordinates of these points of intersection?

9. If the line segment of Ex. 8 is produced beyond the second point until its length is trebled, what will be the coordinates of its extremity?

10. Show that the points (x_1, y_1, z_1), (x_2, y_2, z_2), and (x_3, y_3, z_3) are collinear when there are three numbers k_1, k_2, and k_3, all different from zero, such that

$$k_1 + k_2 + k_3 = 0, \quad k_1x_1 + k_2x_2 + k_3x_3 = 0,$$
$$k_1y_1 + k_2y_2 + k_3y_3 = 0, \quad k_1z_1 + k_2z_2 + k_3z_3 = 0.$$

11. Show that the medians of the triangle with vertices (x_1, y_1, z_1), (x_2, y_2, z_2), and (x_3, y_3, z_3) meet in the point with coordinates $\frac{1}{3}(x_1 + x_2 + x_3)$, $\frac{1}{3}(y_1 + y_2 + y_3)$, $\frac{1}{3}(z_1 + z_2 + z_3)$.

12. Explain why equations (i) of Ex. 5 may be interpreted as giving the coordinates of the center of mass of masses h_2 and h_1 at P_1 and P_2 respectively. Obtain the coordinates of the center of mass of masses m_1, m_2, m_3 at points P_1, P_2, P_3. Do the same for n masses at n different points, using mathematical induction.

16. Equations of a Line. Direction Numbers and Direction Cosines of a Line. Angle of Two Lines

Consider the line through the points $P_1(x_1, y_1, z_1)$ and $P_2(x_2, y_2, z_2)$ and denote by $P(x, y, z)$ a representative point of the line. We consider first the case when the line is not parallel to any one of the coordinate planes, that is, $x_2 \neq x_1$, $y_2 \neq y_1$, and $z_2 \neq z_1$. The segments P_1P and P_1P_2 have the same direction by a characteristic property of a line (see § 6). Consequently their direction numbers are proportional, as follows from (15.5), that is,

$$\frac{x - x_1}{x_2 - x_1} = \frac{y - y_1}{y_2 - y_1} = \frac{z - z_1}{z_2 - z_1}. \tag{16.1}$$

These equations are satisfied by the coordinates x, y, z of any point on the line, and by the coordinates of no other point;

for, if $P(x, y, z)$ is not on the line, the segments P_1P and P_1P_2 do not have the same direction. From (16.1) we obtain the three equations

$$\frac{x - x_1}{x_2 - x_1} - \frac{y - y_1}{y_2 - y_1} = 0, \quad \frac{x - x_1}{x_2 - x_1} - \frac{z - z_1}{z_2 - z_1} = 0, \tag{16.2}$$
$$\frac{y - y_1}{y_2 - y_1} - \frac{z - z_1}{z_2 - z_1} = 0.$$

However, it is readily seen that, if given values of x, y, and z satisfy any two of equations (16.2), they satisfy also the third; that is, two at most of these equations are independent. The geometric significance of equations (16.2) will be shown in § 17.

We consider next the case when the line is parallel to the xy-plane. In this case $z_2 = z_1$, and for any point on the line we have

$$z - z_1 = 0. \tag{16.3}$$

The direction numbers of the segments P_1P and P_1P_2 are $x - x_1$, $y - y_1$, 0 and $x_2 - x_1$, $y_2 - y_1$, 0, and the proportionality of these numbers is expressed by the first of (16.2). Consequently this equation and (16.3) are equations of the line. Similar results hold when the line is parallel to the yz-plane or the xz-plane.

When the line is parallel to the x-axis,

$$y - y_1 = 0, \quad z - z_1 = 0 \tag{16.4}$$

are equations of the line; and similarly for the cases when the line is parallel to the y-axis or the z-axis.

Accordingly we have

[16.1] *Equations* (16.1) *are equations of the line through the points* (x_1, y_1, z_1) *and* (x_2, y_2, z_2), *with the understanding that, if the denominator of any one of the ratios is equal to zero, the numerator equated to zero is one of the equations of the line; these exceptional cases arise when the line is parallel to one of the coordinate planes or to one of the coordinate axes.*

Thus a line, which is one-dimensional, when considered as lying in space, which is three-dimensional, is defined by two equations, whereas when considered as lying in a plane, which is two-dimensional, it is defined by one equation, as in § 5.

Since any two segments of a line have the same direction, their direction numbers are proportional. By definition the direction numbers of any segment of a line are *direction numbers of the line*. Consequently there is an endless number of direction numbers of a line, the numbers of any set being proportional to the corresponding numbers of any other set. Thus, if u, v, w are direction numbers of a line through the point (x_1, y_1, z_1),

$$\frac{x - x_1}{u} = \frac{y - y_1}{v} = \frac{z - z_1}{w} \tag{16.5}$$

are equations of the line, since these equations express the proportionality of two sets of direction numbers of the line for each point (x, y, z) of the line.

Conversely, for each set of numbers u, v, w, not all equal to zero, equations (16.5) are equations of the line through the point (x_1, y_1, z_1) with direction numbers u, v, w. For, if we define numbers x_2, y_2, z_2 by the equations

$$x_2 - x_1 = u, \qquad y_2 - y_1 = v, \qquad z_2 - z_1 = w, \tag{16.6}$$

in terms of x_2, y_2, z_2 equations (16.5) are expressible in the form (16.1), and thus are equations of the line through the points (x_1, y_1, z_1) and (x_2, y_2, z_2). As a result of the above discussion and the fact that any set of direction numbers of a line are direction numbers also of any line parallel to it, we have

[16.2] *If (x_1, y_1, z_1) and (x_2, y_2, z_2) are any two points of a line, the quantities $x_2 - x_1$, $y_2 - y_1$, $z_2 - z_1$ are direction numbers of the line, and of any line parallel to it.*

In defining direction cosines of a line segment in § 15, we assigned sense to a line segment. Since this applies to all segments of a line, we have that the positive sense along a line not parallel to the xy-plane is upward, that is, z increasing; when a line is parallel to the xy-plane and not parallel to the x-axis, the positive sense is the direction in which y increases; and when parallel to the x-axis, the positive sense is the same as on the x-axis.

Since all segments of a line have the same direction, the direction cosines of all segments are the same; we call them the *direction cosines of the line.* Accordingly the direction cosines λ, μ, ν of a line are the cosines of the angles which the positive direction of the line makes with the three line segments from any point on the line and parallel to the positive directions of the x-, y-, and z-axes respectively. Hence Theorem [**15.2**] holds for direction cosines of a line. As a consequence of this theorem we have

[**16.3**] *If u, v, w are direction numbers of a line, the direction cosines of the line are given by*

$$\lambda = \frac{u}{e\sqrt{u^2+v^2+w^2}}, \qquad \mu = \frac{v}{e\sqrt{u^2+v^2+w^2}}, \qquad \nu = \frac{w}{e\sqrt{u^2+v^2+w^2}}, \tag{16.7}$$

where e is $+1$ or -1 so that the first of the numbers ew, ev, and eu which is not zero shall be positive.

In fact, since $e^2 = 1$, the quantities λ, μ, ν given by (16.7) satisfy the conditions of Theorem [**15.2**].

Since equations (15.5) hold for all segments of a line with any particular point (x_1, y_1, z_1) as an end point, we have

[**16.4**] *The line through the point (x_1, y_1, z_1) with direction cosines λ, μ, ν has the parametric equations*

$$x = x_1 + d\lambda, \qquad y = y_1 + d\mu, \qquad z = z_1 + d\nu, \tag{16.8}$$

the parameter d being the directed distance from (x_1, y_1, z_1) to $P(x, y, z)$.

Another set of parametric equations of a line is obtained from equations (16.1) when we observe that, if the line is not parallel to a coordinate plane, for each point (x, y, z) on the line the ratios in (16.1) have the same value, say t, depending upon the values of x, y, and z. If we put each of the ratios in (16.1) equal to t, and solve the resulting equations for x, y, and z, we obtain

$$x = (1-t)x_1 + tx_2, \quad y = (1-t)y_1 + ty_2, \quad z = (1-t)z_1 + tz_2. \tag{16.9}$$

Conversely, the values x, y, z given by (16.9) for any value of t are coordinates of a point on the line with equations (16.1), as one verifies by substitution. Equations (16.9) hold also when the line is parallel to a coordinate plane or coordinate axis, as the reader can show, using a method analogous to that used in connection with equations (5.7). Hence we have

[**16.5**] *Equations* (16.9) *are parametric equations of the line through the points* (x_1, y_1, z_1) *and* (x_2, y_2, z_2); *the coordinates* x, y, z *of any point on the line are given by* (16.9) *for a suitable value of* t, *and conversely.*

Ordinarily two lines in space do not intersect, even if they are not parallel. Two nonintersecting, nonparallel lines are said to be *skew* to one another, and, for the sake of brevity, they are called *skew lines.* If through a point of one of two such lines one draws a line parallel to the other, each of the angles so formed is called an *angle of the two skew lines.* Accordingly from Theorem [**15.3**] we have

[**16.6**] *The angle* ϕ *of the positive directions of two lines with direction cosines* λ_1, μ_1, ν_1 *and* λ_2, μ_2, ν_2 *is given by*

$$\cos \phi = \lambda_1\lambda_2 + \mu_1\mu_2 + \nu_1\nu_2. \tag{16.10}$$

From this theorem and Theorem [**16.3**] we have

[**16.7**] *The angle* ϕ *of the positive directions of two lines with direction numbers* u_1, v_1, w_1 *and* u_2, v_2, w_2 *is given by*

$$\cos \phi = \frac{u_1u_2 + v_1v_2 + w_1w_2}{e_1e_2\sqrt{(u_1^2 + v_1^2 + w_1^2)(u_2^2 + v_2^2 + w_2^2)}}, \tag{16.11}$$

where e_1 *is* $+1$ *or* -1 *so that the first of the numbers* e_1w_1, e_1v_1, *and* e_1u_1 *which is not zero shall be positive, and similarly for* e_2.

As a corollary we have

[**16.8**] *Two lines with direction numbers* u_1, v_1, w_1 *and* u_2, v_2, w_2 *are perpendicular, if and only if*

$$u_1u_2 + v_1v_2 + w_1w_2 = 0. \tag{16.12}$$

EXERCISES

1. Find the distance between the points $(2, -1, 3)$ and $(-3, 2, 5)$, and the direction cosines of the line through these points.

2. Find equations of the line through the points $(2, 1, -3)$ and $(3, -2, 1)$; through the points $(1, 3, -2)$ and $(-2, 3, 1)$.

3. For what value of a do the points $(a, -3, 10)$, $(2, -2, 3)$, and $(6, -1, -4)$ lie on a line?

4. What are the direction cosines of the coordinate axes? What are equations of the axes? What are the direction cosines of the lines through the origin bisecting the angles between the coordinate axes? What are equations of these lines?

5. Find the dirnction cosines of the lines equally inclined to the coordinate axes.

6. Direction numbers of a line through the point $(2, -1, 3)$ are $3, 1, -4$; find the equations of the line in the parametric form (16.8).

7. Show by means of (16.9) that if the coordinates of two points (x_1, y_1, z_1) and (x_2, y_2, z_2) are in the relation

$$x_2 = kx_1, \quad y_2 = ky_1, \quad z_2 = kz_1,$$

where k is some constant different from zero, the line joining the two points passes through the origin. How does this follow from geometric considerations without the use of (16.9)?

8. Find the cosine of the angle of the positive directions of the lines through the points $(3, -1, 0)$, $(1, 2, 1)$ and $(-2, 0, 1)$, $(1, 2, 0)$ respectively.

9. For what value of a are lines with direction numbers $1, -2, 2$ and $2, 2, a$ perpendicular? For what value of a is the cosine of the angle of the positive directions of these lines equal to $4/9$?

10. Find the direction cosines of a line perpendicular to each of two lines whose respective direction numbers are $3, 5, 6$ and $1, 3, 4$.

11. Show that equations (16.5) may be written

$$x = x_1 + ul, \quad y = y_1 + vl, \quad z = z_1 + wl,$$

where l is a parameter. What relation does l bear to the directed distance from (x_1, y_1, z_1) to (x, y, z)?

12. Show that equations (i) of Ex. 5, § 15 are parametric equations of the line through (x_1, y_1, z_1) and (x_2, y_2, z_2), and find the relation between h_1 and h_2 in these equations and t in (16.9).

13. Show that the parameter t in (16.9) is equal to d/d_2, where d and d_2 are the directed distances of the points (x, y, z) and (x_2, y_2, z_2) respectively from the point (x_1, y_1, z_1).

14. Find the locus of a point equidistant from the three coordinate planes.

15. Find the locus of a point whose distances from the xy-, yz-, and xz-planes are in the ratio $1 : 2 : 3$.

16. Find the locus of a point $P(x, y, z)$ so that the line joining P to $P_1(1, -2, 3)$ is perpendicular to the line through P_1 with direction numbers $2, 1, -3$.

17. Find the locus of a point at the distance 2 from the point $(1, -2, 3)$.

17. An Equation of a Plane

In defining coordinates in § 14 we took it for granted that the reader understood what is meant by a plane, and that when two planes meet they intersect in a straight line. Euclid proved the latter result by means of his definition of a plane as a surface such that a straight line joining any two points of the surface lies entirely in the surface, as stated in § 14. By means of this characteristic property of the plane we shall prove the theorem

[17.1] *Any equation of the first degree in x, y, and z is an equation of a plane.*

Consider the equation

$$ax + by + cz + d = 0. \tag{17.1}$$

It is understood that the coefficient of at least one of the unknowns is different from zero, that is, that we are dealing with a nondegenerate equation (see § 1); this understanding applies to all theorems concerning equations of the first degree in x, y, z in this chapter.

Let $P_1(x_1, y_1, z_1)$ and $P_2(x_2, y_2, z_2)$ be any two points of the locus defined by equation (17.1); then we have

$$ax_1 + by_1 + cz_1 + d = 0, \quad ax_2 + by_2 + cz_2 + d = 0. \tag{17.2}$$

In § 16 it was shown that the coordinates x, y, z of any point of the line through the points P_1 and P_2 are given by (16.9) for an appropriate value of the parameter t. When the expressions (16.9) are substituted in the left-hand member of (17.1), the resulting expression may be written in the form

$$(1-t)(ax_1+by_1+cz_1+d)+t(ax_2+by_2+cz_2+d).$$

In consequence of (17.2) this expression is equal to zero for every value of t. Hence the coordinates of every point of the line through P_1 and P_2 satisfy (17.1); that is, every point of this line is a point of the locus defined by (17.1). Since this result holds for every pair of points P_1 and P_2 whose coordinates satisfy (17.1), the theorem follows from Euclid's definition of a plane.

We shall consider several particular forms of equation (17.1) and in the first place prove the theorem

[17.2] *When two and only two of the coefficients of x, y, and z in equation* (17.1) *are equal to zero, the locus is one of the coordinate planes or a plane parallel to it according as $d=0$ or $d\neq 0$.*

Consider, for example, the case when $a=b=0$, that is, the equation

$$0\,x+0\,y+cz+d=0 \text{ (usually written } cz+d=0). \tag{17.3}$$

This equation is satisfied by $z=-d/c$ and any values of x and y; consequently it is an equation of the xy-plane when $d=0$, and when $d\neq 0$ of the plane parallel to the xy-plane and at the distance $-d/c$ from it, above or below it according as $-d/c$ is positive or negative. In like manner,

$$ax+0\,y+0\,z+d=0, \qquad 0\,x+by+0\,z+d=0 \tag{17.4}$$

are equations respectively of the yz-plane or a plane parallel to it and of the xz-plane or a plane parallel to it according as $d=0$ or $d\neq 0$.

Next we prove the theorem

[17.3] *When one and only one of the coefficients of x, y, and z in equation* (17.1) *is equal to zero, the locus is a plane through one of the coordinate axes or parallel to it according as $d=0$ or $d\neq 0$.*

Consider, for example, the case when $c = 0$, that is, the equation

(17.5) $ax + by + 0\,z + d = 0$ (usually written $ax + by + d = 0$).

The points $(x, y, 0)$, where x and y are solutions of this equation, lie upon a line in the xy-plane, namely, the line whose equation in two dimensions, as discussed in Chapter 1, is $ax + by + d = 0$. But any such pair x, y, say x_1 and y_1, and any z satisfy equation (17.5) in three dimensions; hence the point (x_1, y_1, z) for any z is on the line through $(x_1, y_1, 0)$ parallel to the z-axis. Therefore equation (17.5) is an equation of a plane perpendicular to the xy-plane and meeting the latter in the line whose equation in two dimensions is $ax + by + d = 0$; when $d = 0$, this plane passes through the z-axis.

In like manner

(17.6) $0\,x + by + cz + d = 0$ (usually written $by + cz + d = 0$)

and

(17.7) $ax + 0\,y + cz + d = 0$ (usually written $ax + cz + d = 0$)

are equations of planes perpendicular to the yz-plane and the xz-plane respectively.

Referring now to equations (16.2), obtained from equations (16.1) of the line through the points (x_1, y_1, z_1), (x_2, y_2, z_2), we observe that these equations are equations of planes, each containing the line, and parallel to the z-, y-, and x-axes respectively, as shown in Fig. 19.

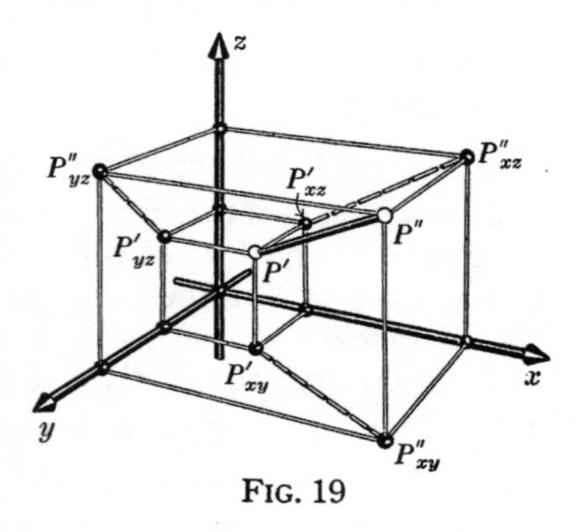

FIG. 19

Next we shall prove the converse of Theorem **[17.1]**, namely,

[17.4] *Any plane is defined algebraically by an equation of the first degree in x, y, and z.*

In proving this theorem we remark that the xy-plane is defined by (17.3) for $d = 0$, that is, by $z = 0$, and that any plane parallel to the xy-plane is defined by (17.3) with c and d numbers such that $-d/c$ is equal to the directed distance of the plane from the xy-plane. Any other plane intersects the xy-plane in a line. In deriving an equation of such a plane we make use of the property of a plane that it is completely determined by three noncollinear points, that is, by three points not on the same line, or, what is the same thing, by a line and a point not on the line; this property follows from Euclid's definition of a plane, stated at the beginning of this section. The line in which the given plane cuts the xy-plane is defined in this plane by an equation $ax + by + d = 0$ (in accordance with the results of § 5) when we are dealing with the geometry of the plane; but when we are dealing with the line in space, it is defined by this equation and the equation $z = 0$. Consider then the equation

$$ax + by + d + cz = 0, \tag{17.8}$$

where a, b, and d have the values from the above equation of the line in the xy-plane, and c is as yet unassigned. Whatever be c, by Theorem [**17.1**] this is an equation of a plane, in fact, of a plane through the line, since the coordinates of any point on the line satisfy $ax + by + d = 0$ and $z = 0$, and consequently (17.8). Suppose now that (x_1, y_1, z_1) is a point of the given plane not on its line of intersection with the xy-plane; then $z_1 \neq 0$. If now c is found from the equation

$$ax_1 + by_1 + d + cz_1 = 0, \tag{17.9}$$

and this value is substituted in (17.8), the resulting equation is an equation of the given plane, since the plane with this equation passes through the point (x_1, y_1, z_1), in consequence of (17.9), and through the line $ax + by + d = 0$ in the plane $z = 0$; and the theorem is proved.

From Theorem [**17.1**] and equations (17.1) and (17.2) it follows that

$$a(x - x_1) + b(y - y_1) + c(z - z_1) = 0 \tag{17.10}$$

is an equation of a plane containing the point $P_1(x_1, y_1, z_1)$. Since $x - x_1$, $y - y_1$, $z - z_1$ are direction numbers of the line joining P_1 to any point (x, y, z) in the plane, it follows from (17.10) and Theorem [**16.8**] that a, b, c are direction numbers of a line perpendicular to every line in the plane which passes through P_1; hence the line through P_1 with direction numbers a, b, c is perpendicular to the plane at P_1. We follow the custom of saying that this line is the *normal* to the plane at P_1. Since P_1 is any point of the plane, a, b, c are direction numbers of every normal to the plane; evidently all the normals are parallel. Hence we have (a generalization of Theorem [**6.9**])

[**17.5**] *The geometric significance of the coefficients a, b, c of an equation $ax + by + cz + d = 0$ of a plane is that they are direction numbers of each of the normals to the plane.*

From Theorem [**17.5**] and equations (16.5) we have

[**17.6**] *The equation*

$$a(x - x_1) + b(y - y_1) + c(z - z_1) = 0$$

is an equation of a plane through the point (x_1, y_1, z_1), *and*

$$\frac{x - x_1}{a} = \frac{y - y_1}{b} = \frac{z - z_1}{c} \tag{17.11}$$

are equations of the normal to the plane at this point.

Compare this theorem and Theorem [**6.10**].

If a line lies in a plane, or is parallel to a plane, it is perpendicular to the normals to the plane, and conversely. Hence by Theorem [**17.5**] we have

[**17.7**] *A line with direction numbers u, v, w is parallel to, or lies in, the plane with equation* (17.1), *if and only if*

$$au + bv + cw = 0. \tag{17.12}$$

When two planes are perpendicular, the angle of the normals to the planes at a point on their line of intersection is a right angle, and conversely. Hence in consequence of Theorems [**16.8**] and [**17.5**] we have

[**17.8**] *The planes with equations*

$$a_1x + b_1y + c_1z + d_1 = 0, \qquad a_2x + b_2y + c_2z + d_2 = 0 \tag{17.13}$$

are perpendicular to one another, if and only if

$$a_1a_2 + b_1b_2 + c_1c_2 = 0. \tag{17.14}$$

If we wish to construct a plane through a point P parallel to a given plane, we draw through P two lines parallel to the given plane, and the plane determined by these two lines is the plane desired. This construction is equivalent to that of constructing the plane through P perpendicular to the line through P which is normal to the given plane. Hence a line normal to one of two parallel planes is normal to the other also. In consequence of this result and of Theorem [**17.5**] we have

[**17.9**] *An equation of any plane parallel to the plane*

$$ax + by + cz + d = 0$$

is

$$ax + by + cz + e = 0 \tag{17.15}$$

for a suitable value of the coefficient e.

We have remarked that a plane is determined by three noncollinear points. Accordingly, if we wish to find an equation of a plane through three noncollinear points, we have only to substitute the coordinates of the points in equation (17.1) and solve the three equations thus obtained for a, b, and c. When the resulting expressions are substituted in (17.1), d appears as a factor, unless $d = 0$, which emphasizes the fact that an equation of the first degree in x, y, and z, and any constant multiple of this equation, are equations of the same plane. More expeditious methods of finding an equation of a plane when the coordinates of three of its points are given are developed in §§ 21 and 23.

EXERCISES

1. Find an equation of the plane through the points (2, 3, 0) and (− 2, 1, 0) parallel to the z-axis.

2. Find an equation of the plane through the three points (1, 1, 1), (− 1, 1, 1), and (− 3, − 7, − 5). Is the plane parallel to one of the axes?

3. Find an equation of the plane through the origin and the points (1, 3, 2) and (2, − 1, − 1).

4. Show that the points (1, 1, 1), (1, 2, − 1), and (1, 3, − 3) are collinear, and find an equation of the set of planes containing these points.

5. Find the intercepts on the coordinate axes of the plane

$$\frac{x}{g}+\frac{y}{h}+\frac{z}{k}=1.$$

When can an equation of a plane not be put in this *intercept form*?

6. Find the locus of a point whose distance from the xy-plane is twice its distance from the yz-plane.

7. For what value of the coefficient a are the planes

$$ax-2y+z+7=0 \quad \text{and} \quad 3x+4y-2z+1=0$$

perpendicular to one another?

8. Find equations of two planes through the point (− 1, 2, 0) perpendicular to the plane

$$\text{(i)} \qquad 2x-2y+5z-10=0,$$

one perpendicular to the xy-plane and the other perpendicular to the yz-plane. Using this result, find an equation of the set of planes through the point perpendicular to the plane (i).

9. Find an equation of the plane through the line

$$\frac{x-1}{2}=\frac{y+2}{-3}=\frac{z}{5}$$

perpendicular to the plane $x-y+z+2=0$.

10. Find an equation of the plane parallel to the plane

$$7x-3y+z-5=0$$

and containing the point (1, − 2, 3).

11. For what value of a is the plane $3x-2y+az=0$ perpendicular to the plane $3x+4y-7z+1=0$?

12. Find an equation of the locus of a point equidistant from the points (8, 3, 4) and (3, − 1, − 2), and show that it is the plane perpendicular to the line segment joining the two points at the midpoint of the segment.

13. Show that the angle ϕ of the positive directions of the normals to the planes with equations (17.13) is given by

$$\cos\phi=\frac{a_1a_2+b_1b_2+c_1c_2}{e_1e_2\sqrt{(a_1{}^2+b_1{}^2+c_1{}^2)(a_2{}^2+b_2{}^2+c_2{}^2)}},$$

and state under what conditions e_1 is $+1$ or -1, and e_2 is $+1$ or -1.

14. Show that when the expressions (i) of Ex. 5, § 15 are substituted in an equation of a plane, the resulting value of the ratio h_1/h_2 gives the ratio in which the segment P_1P_2, or the segment produced, is divided by the point of intersection of the line of the segment and the given plane. Find the ratio in which the line segment joining the points $(2, 3, -1)$ and $(5, -6, 2)$ is divided by the point of intersection of the line of the segment and the plane $3x - 2y + 6z - 2 = 0$; find also the coordinates of the point of intersection.

18. The Directed Distance from a Plane to a Point. The Distance from a Line to a Point

We shall establish the following theorem:

[18.1] *The directed distance from the plane*

$$ax + by + cz + d = 0 \tag{18.1}$$

to the point (x_1, y_1, z_1) *is given by*

$$l = \frac{ax_1 + by_1 + cz_1 + d}{e\sqrt{a^2 + b^2 + c^2}}, \tag{18.2}$$

where e *is* $+1$ *or* -1 *so that the first of the numbers* ec, eb, ea *which is not zero shall be positive.*

When $b = c = 0$, equation (18.1) of the plane reduces to $x + \frac{d}{a} = 0$; that is, the plane is parallel to the yz-plane and at the distance $-d/a$ from this plane. Consequently the distance from the plane to P_1 is equal to $x_1 - \left(-\frac{d}{a}\right) = x_1 + \frac{d}{a}$, which result follows from (18.2). In this case l is positive or negative according as P_1 lies to the right or left of the given plane.

We consider next the case when $c = 0$, in which case the plane is perpendicular to the xy-plane and passes through the line $ax + by + d = 0$, $z = 0$. It is evident that the distance from the plane to $P_1(x_1, y_1, z_1)$ is equal to the distance in the xy-plane from the line $ax + by + d = 0$ to the point $(x_1, y_1, 0)$. But as given by (8.5) this distance is the number obtained from (18.2) on putting $c = 0$.

We consider finally the case when $c \neq 0$, that is, when the plane cuts the z-axis. From Theorems [**17.5**] and [**16.3**] we have that the direction cosines of a normal to the plane are given by

$$\lambda, \mu, \nu = \frac{a, b, c}{e\sqrt{a^2 + b^2 + c^2}}, \tag{18.3}$$

where e is $+1$ or -1 according as $c > 0$ or $c < 0$. If we denote by $P_2(x_2, y_2, z_2)$ the point in which the normal to the plane through P_1 meets the plane, we have from (15.5)

$$x_1 - x_2 = l\lambda, \qquad y_1 - y_2 = l\mu, \qquad z_1 - z_2 = l\nu, \tag{18.4}$$

where l is the distance from P_2 to P_1, positive or negative according as P_1 is above or below P_2 on the normal. Solving equations (18.4) for x_2, y_2, and z_2, and substituting the resulting expressions in (18.1), we obtain, on rearranging the terms and making use of (18.3),

$$\begin{aligned} ax_1 + by_1 + cz_1 + d = l(a\lambda + b\mu + c\nu) &= l\frac{a^2 + b^2 + c^2}{e\sqrt{a^2 + b^2 + c^2}} \\ &= le\sqrt{a^2 + b^2 + c^2}, \end{aligned}$$

since $1/e = e$, from which we derive the expression (18.2); and the theorem is proved.

We turn now to the problem of finding the distance d from a line to a point $P_1(x_1, y_1, z_1)$. Let λ, μ, ν be direction cosines of the line, and $P_2(x_2, y_2, z_2)$ a point on the line. Denote by θ the angle between the line and the line segment P_2P_1; then

$$d = \overline{P_1P_2} \sin \theta, \tag{18.5}$$

where $\overline{P_1P_2}$ denotes the length of the line segment. The direction cosines of the line segment P_1P_2 are, to within sign at most,

$$\frac{x_2 - x_1, \quad y_2 - y_1, \quad z_2 - z_1}{\overline{P_1P_2}}.$$

From these expressions, (18.5), and (15.10) we have as the desired result

$$\begin{aligned} d^2 = \begin{vmatrix} y_2 - y_1 & z_2 - z_1 \\ \mu & \nu \end{vmatrix}^2 + \begin{vmatrix} z_2 - z_1 & x_2 - x_1 \\ \nu & \lambda \end{vmatrix}^2 \\ + \begin{vmatrix} x_2 - x_1 & y_2 - y_1 \\ \lambda & \mu \end{vmatrix}^2. \end{aligned} \tag{18.6}$$

EXERCISES

1. Find the distance from the plane $2x - y + 2z - 6 = 0$ to the point $(1, -1, 2)$, and the coordinates of the point in which the normal to the plane from the point meets the plane.

2. Find an equation of the plane through the point $(-1, 2, 3)$ parallel to the plane $x + 2y - 2z + 3 = 0$. What is the distance between the planes?

3. Find equations of the two planes at the distance 5 from the origin and perpendicular to a line with direction numbers $1, -3, 2$.

4. What is the distance between the planes $2x - y + z - 5 = 0$ and $4x - 2y + 2z + 3 = 0$?

5. Given two intersecting planes and a point P in one of the four compartments into which the planes divide space, that is, one of the four *dihedral angles* of the planes, show that the dihedral angle in which P lies, and the angle ϕ between the positive directions of the normals to the two planes through P are supplementary when P lies on the positive or negative sides of the two planes, and that these angles are equal when P is on the positive side of one plane and the negative side of the other.

6. Determine whether the origin lies in an acute or obtuse dihedral angle between the planes $2x + 3y - 6z + 3 = 0$ and $8x - y + 4z - 5 = 0$.

7. Find equations of the planes which are parallel to the plane $2x + 2y - z + 6 = 0$ and at the distance 2 from it (see equation (8.6)).

8. Find the locus of a point equidistant from the planes

$$a_1x + b_1y + c_1z + d_1 = 0 \quad \text{and} \quad a_2x + b_2y + c_2z + d_2 = 0,$$

and show that it consists of the two planes which bisect the angles formed by the given planes when the latter intersect; discuss also the case when the given planes are parallel. Apply the result to the planes in Ex. 6.

9. Find the locus of a point which is twice as far from the plane $2x + 2y - z + 3 = 0$ as from the plane $x - 2y + 2z - 6 = 0$.

10. Show that for the plane containing the two lines

$$\frac{x}{u_1} = \frac{y}{v_1} = \frac{z}{w_1}, \qquad \frac{x}{u_2} = \frac{y}{v_2} = \frac{z}{w_2},$$

lines with direction numbers $k_1u_1 + k_2u_2$, $k_1v_1 + k_2v_2$, $k_1w_1 + k_2w_2$ for any values of the constants k_1 and k_2, not both zero, lie in this plane or are parallel to it.

11. Find the angle which the line $\frac{x}{3} = \frac{y}{2} = \frac{z}{-1}$ makes with its projection on the plane $x + 2y - 3z = 0$. This angle is called the *angle the line makes with the plane.*

12. Find what relation or relations must hold in each case between the coefficients of the equation $ax + by + cz + d = 0$ of a plane in order that the plane

(*a*) have the intercept 2 on the y-axis;
(*b*) have equal intercepts;
(*c*) be parallel to the plane $2x - 3y + 2z - 1 = 0$;
(*d*) be perpendicular to the yz-plane;
(*e*) contain the point $(5, -4, 2)$;
(*f*) contain the line $\frac{x}{2} = \frac{y}{-1} = \frac{z}{3}$;
(*g*) be parallel to the x-axis;
(*h*) be at the distance $+2$ from the origin.

13. Show that if (x_1, y_1, z_1) and (x_2, y_2, z_2) are points of two parallel planes with equations $ax + by + cz + d_1 = 0$ and $ax + by + cz + d_2 = 0$, the distance between the planes is the numerical value of

$$\text{(i)} \qquad \frac{a(x_2 - x_1) + b(y_2 - y_1) + c(z_2 - z_1)}{\sqrt{a^2 + b^2 + c^2}}.$$

Under what conditions is this expression a positive number?

14. Show that the numerator in the expression (i) of Ex. 13 is equal to $d_1 - d_2$, and that the number given by the expression (i) is equal to the length of the segment between the two planes of any normal (the end points of the segment being the projections upon the normal of any point in each plane).

19. Two Equations of the First Degree in Three Unknowns. A Line as the Intersection of Two Planes

Consider the two equations

$$(19.1) \qquad a_1x + b_1y + c_1z + d_1 = 0, \qquad a_2x + b_2y + c_2z + d_2 = 0.$$

Since by hypothesis the coefficient of at least one of the unknowns in each equation is different from zero, two of the unknowns in either equation can be given arbitrary values, and then the other can be found. Thus, when $a_1 \neq 0$, if we give

y and z any values in the first of equations (19.1), and solve for the value of x, this value and the given values of y and z constitute a solution of the equation. Although each of equations (19.1) admits an endless number of solutions involving two arbitrary choices, it does not follow necessarily that the equations have a common solution.

We assume that equations (19.1) have a common solution. Instead of denoting it by x_1, y_1, z_1 to make it evident that we are dealing with a particular solution (after the manner followed in § 9), we count on the reader's thinking of x, y, and z in what follows as the same set of numbers in the two equations. If we multiply the first of equations (19.1) by b_2 and from the result subtract the second of (19.1) multiplied by b_1, the final result may be written

$$(19.2)\qquad \begin{vmatrix} a_1 & b_1 \\ a_2 & b_2 \end{vmatrix} x - \begin{vmatrix} b_1 & c_1 \\ b_2 & c_2 \end{vmatrix} z + \begin{vmatrix} d_1 & b_1 \\ d_2 & b_2 \end{vmatrix} = 0.$$

This is the process followed in § 9, leading to equation (9.4); we refer to it as eliminating y from the two equations, meaning that from these equations we obtain an equation in which the coefficient of y is zero. If, in similar manner, we eliminate x and z respectively from equations (19.1), we obtain

$$(19.3)\qquad \begin{vmatrix} a_1 & b_1 \\ a_2 & b_2 \end{vmatrix} y - \begin{vmatrix} c_1 & a_1 \\ c_2 & a_2 \end{vmatrix} z + \begin{vmatrix} a_1 & d_1 \\ a_2 & d_2 \end{vmatrix} = 0$$

and

$$(19.4)\qquad \begin{vmatrix} c_1 & a_1 \\ c_2 & a_2 \end{vmatrix} x - \begin{vmatrix} b_1 & c_1 \\ b_2 & c_2 \end{vmatrix} y + \begin{vmatrix} c_1 & d_1 \\ c_2 & d_2 \end{vmatrix} = 0.$$

If the determinant $\begin{vmatrix} a_1 & b_1 \\ a_2 & b_2 \end{vmatrix}$ is not equal to zero, when z in equations (19.2) and (19.3) is given any value, these equations can be solved for x and y, which values together with that assigned to z constitute a common solution of equations (19.1). Thus in this case there is an endless number of common solutions of these equations. In like manner, if $\begin{vmatrix} b_1 & c_1 \\ b_2 & c_2 \end{vmatrix}$ is not equal to zero, equations (19.2) and (19.4) can be solved for y and z

for any given value of x; and if $\begin{vmatrix} c_1 & a_1 \\ c_2 & a_2 \end{vmatrix}$ is not equal to zero, equations (19.3) and (19.4) can be solved for x and z for any given value of y. Hence we have

[19.1] *Equations* (19.1) *have an endless number of solutions when any one of the determinants*

$$(19.5)\qquad \begin{vmatrix} a_1 & b_1 \\ a_2 & b_2 \end{vmatrix}, \qquad \begin{vmatrix} b_1 & c_1 \\ b_2 & c_2 \end{vmatrix}, \qquad \begin{vmatrix} c_1 & a_1 \\ c_2 & a_2 \end{vmatrix}$$

is not equal to zero.

When one at least of the determinants (19.5) is not equal to zero, the common solutions of equations (19.1) are coordinates of points of the line of intersection of the two planes with the equations (19.1). Hence we have

[19.2] *Two equations* (19.1) *are equations of a line when one at least of the determinants* (19.5) *is not equal to zero; this line is the intersection of the planes with equations* (19.1).

In § 20 it will be shown how equations (19.1) can be put in the form (16.5) when a common solution is known.

We turn now to the consideration of the case when all the determinants (19.5) are equal to zero. By Theorem [**9.6**] this condition is equivalent to the three equations

$$(19.6)\qquad a_2 = ta_1, \qquad b_2 = tb_1, \qquad c_2 = tc_1,$$

where t is some number, not zero. By means of (19.6) the last determinants in equations (19.2), (19.3), and (19.4) reduce to the respective values

$$-b_1(d_2 - td_1), \qquad a_1(d_2 - td_1), \qquad c_1(d_2 - td_1).$$

If not all three of these quantities are equal to zero, one at least of (19.2), (19.3), and (19.4) cannot be true when all three of the determinants (19.5) are equal to zero, and consequently the assumption that equations (19.1) have a common solution is false. If $d_2 \neq td_1$, all three of the above expressions can be equal to zero only in case $a_1 = b_1 = c_1 = 0$, which is contrary to the hypothesis that one at least of these coefficients is not zero. If, however, $d_2 = td_1$, all the above expressions are equal to zero. But the equations $d_2 = td_1$ and (19.6) are the conditions that the second of (19.1) is a constant multiple of the first. Hence we have

[**19.3**] *Two equations of the first degree in three unknowns neither of which is a constant multiple of the other admit an endless number of common solutions or none; the condition for the latter in terms of equations* (19.1) *is*

$$\frac{a_2}{a_1} = \frac{b_2}{b_1} = \frac{c_2}{c_1} \neq \frac{d_2}{d_1}, \tag{19.7}$$

with the understanding that if one a is zero so is the other a, and similarly for the b's and c's.

If we say that two planes are parallel when they do not have a point in common, Theorem [**19.3**] may be stated as follows:

[**19.4**] *The planes with equations* (19.1) *are parallel, if and only if*

$$\frac{a_2}{a_1} = \frac{b_2}{b_1} = \frac{c_2}{c_1} \neq \frac{d_2}{d_1}.$$

Another way of stating this theorem is

[**19.5**] *A plane is parallel to the plane* $ax + by + cz + d = 0$, *if and only if its equation is* $ax + by + cz + e = 0$ *where* $e \neq d$, *or any constant multiple of this equation.*

This theorem is equivalent to Theorem [**17.9**], which was derived from another, and consequently an equivalent, definition of parallel planes.

Next we establish the following theorem, which is a generalization of Theorem [**10.1**]:

[**19.6**] *When two equations* (19.1) *are equations of a line, that is, when the coefficients are not proportional, the equation*

$$t_1(a_1x + b_1y + c_1z + d_1) + t_2(a_2x + b_2y + c_2z + d_2) = 0, \tag{19.8}$$

for any values of the constants t_1 *and* t_2, *not both zero, is an equation of a plane through the line defined by* (19.1); *and* (19.8) *is an equation of each plane through this line for suitable values of* t_1 *and* t_2.

In order to prove the theorem, we rewrite equation (19.8) in the form

$$(19.9) \qquad (a_1t_1 + a_2t_2)x + (b_1t_1 + b_2t_2)y + (c_1t_1 + c_2t_2)z + (d_1t_1 + d_2t_2) = 0.$$

Since by hypothesis the coefficients of x, y, z in (19.1) are not proportional, it is impossible to find values of t_1 and t_2, not both zero, which will make all the coefficients of x, y, z in (19.9) equal to zero. Consequently from Theorem [17.1] we have that equation (19.9) is an equation of a plane for any values of t_1 and t_2, not both zero. Moreover, it is an equation of a plane through the line with equations (19.1); for, the two expressions in the parentheses in (19.8) reduce to zero when x, y, and z are given the values of the coordinates of any point on the line. Hence the coordinates of every point on the line satisfy equation (19.8) whatever be t_1 and t_2; and consequently for each choice of t_1 and t_2 equation (19.8) is an equation of a plane containing the line.

Conversely, any particular plane containing the line is determined by any point of the plane not on the line, say (x_1, y_1, z_1). If the values x_1, y_1, z_1 are substituted in (19.8), we obtain

$$(19.10) \qquad t_1A_1 + t_2A_2 = 0,$$

where A_1 and A_2 are the numbers to which the expressions in parentheses in (19.8) reduce when x, y, z are replaced by x_1, y_1, z_1. Moreover, not both A_1 and A_2 are equal to zero, since the point (x_1, y_1, z_1) is not on the line. If the point (x_1, y_1, z_1) is on the first of the planes (19.1), then $A_1 = 0$, and from (19.10) it follows that $t_2 = 0$ and that t_1 can take any value. Also, if (x_1, y_1, z_1) is on the second of the planes (19.1), we have $A_2 = 0$ and $t_1 = 0$. For any other plane both A_1 and A_2 are different from zero. If then we choose any value other than zero for t_1, substitute this value in (19.10), and solve for t_2, the resulting value of t_2 and the chosen value of t_1 are such that when they are substituted in (19.8) the resulting equation is an equation of the plane through the line with equations (19.1) and the point (x_1, y_1, z_1). Thus the theorem is proved.

If we take two sets of values of t_1 and t_2 which are not proportional, the corresponding equations (19.8) are equations of two different planes, and these two equations are equations of the line (19.1). Hence we have

[19.7] *Equations of any two of the set of all the planes through a line constitute equations of the line.*

EXERCISES

1. Find an equation of the first degree in x, y, and z which has $(1, -2, 3)$ for a solution but which has no solution in common with the equation $5x + 2y - 3z + 1 = 0$.

2. Find an equation of the plane through the point $(3, -1, 2)$ and parallel to the plane $x - 2y + 7z + 2 = 0$.

3. Find equations of the lines in which the three coordinate planes are cut by the plane $3x - 4y + 5z - 10 = 0$.

4. Find an equation of the plane containing the line

$$2x - 3y + z + 2 = 0, \qquad 3x + 2y - z + 2 = 0$$

and the point $(1, -1, 1)$; containing the line and parallel to the z-axis.

5. Find an equation of the plane through the line

$$3x - 2y - z - 3 = 0, \qquad 2x + y + 4z + 1 = 0$$

which makes equal intercepts on the x- and y-axes.

6. What is the locus of the equation $(ax + by + cz + d)^2 = k^2$, when $k = 0$; when $k \neq 0$?

7. Show that the plane $x + 2z = 0$ is parallel to a plane containing the line

$$x - 2y + 4z + 4 = 0, \qquad x + y + z - 8 = 0,$$

and consequently is parallel to the line.

8. When the planes (19.1) are parallel, for what values of t_1 and t_2 is equation (19.8) an equation of a plane parallel to the planes (19.1)?

9. Find the condition upon t_1 and t_2 in equation (19.8) so that it is an equation of a plane parallel to the x-axis; to the y-axis; to the z-axis.

10. Show that, when (19.1) are equations of a line, equations (19.2), (19.3), and (19.4) are equations of planes through the line parallel to the y-, x-, and z-axes respectively.

20. Two Homogeneous Equations of the First Degree in Three Unknowns

We consider next the particular case of equations (19.1) when $d_1 = d_2 = 0$; that is, we consider the equations

$$a_1x + b_1y + c_1z = 0, \qquad a_2x + b_2y + c_2z = 0, \tag{20.1}$$

with the understanding that they are independent, that is, that not all of the determinants (19.5) are equal to zero. We say that each of equations (20.1) is *homogeneous* of the first degree in the unknowns, since every term is of the first degree in these unknowns. On putting $d_1 = d_2 = 0$ in (19.2), (19.3), and (19.4), we have

$$\begin{gathered} |a_1b_2|\,x = |b_1c_2|\,z, \qquad |a_1b_2|\,y = -|a_1c_2|\,z, \\ |b_1c_2|\,y = -|a_1c_2|\,x, \end{gathered} \tag{20.2}$$

where for the sake of brevity we have indicated only the elements of the *main diagonal*; that is, $|a_1b_2|$ stands for

$$\begin{vmatrix} a_1 & b_1 \\ a_2 & b_2 \end{vmatrix}.$$

We have also made use of the fact that $|c_1a_2| = -|a_1c_2|$. These equations are satisfied by $x = y = z = 0$, which is an evident common solution of equations (20.1). But they are satisfied also by

$$x = t\,|b_1c_2|, \qquad y = -t\,|a_1c_2|, \qquad z = t\,|a_1b_2| \tag{20.3}$$

for any value of t. Hence we have

[20.1] *Two independent homogeneous equations of the first degree in three unknowns admit an endless number of common solutions; for equations* (20.1) *the solutions are given by* (20.3), *which may be written in the form*

$$x : y : z = |b_1c_2| : -|a_1c_2| : |a_1b_2|, \tag{20.4}$$

with the understanding that if any one of the determinants is zero the corresponding unknown is zero for all common solutions.

The reader should observe that (20.3) follows from (20.4), just as (20.4) follows from (20.3).

We return to the consideration of equations (19.1) when one at least of the determinants (19.5) is not equal to zero, and denote by x_1, y_1, z_1 a common solution of these equations. Then equations (19.1) can be written in the form

$$a_1(x - x_1) + b_1(y - y_1) + c_1(z - z_1) = 0,$$
$$a_2(x - x_1) + b_2(y - y_1) + c_2(z - z_1) = 0.$$

These equations being homogeneous of the first degree in $x - x_1$, $y - y_1$, and $z - z_1$, it follows from Theorem [20.1] that

$$\frac{x - x_1}{|b_1c_2|} = \frac{y - y_1}{-|a_1c_2|} = \frac{z - z_1}{|a_1b_2|}. \tag{20.5}$$

These equations being of the form (16.5), we have

[20.2] *Two equations* (19.1), *for which one at least of the determinants*

$$|b_1c_2|, \qquad -|a_1c_2|, \qquad |a_1b_2| \tag{20.6}$$

is different from zero, are equations of a line with these determinants as direction numbers.

EXERCISES

1. Find the common solutions of the equations

$$x - 2y + 4z = 0, \qquad 2x + y - z = 0.$$

2. Verify by substitution that (20.3) is a common solution of equations (20.1) for every value of t.

3. Show that the point $(1, -2, 0)$ is on the line

$$2x + 3y - 2z + 4 = 0, \qquad 2x - 3y - 5z - 8 = 0,$$

and derive equations of the line in the form (16.5).

4. Find equations of the line parallel to the line in Ex. 3 and through the point $(2, 1, -1)$.

5. Find an equation of the plane through the origin and perpendicular to the line

$$4x - y + 3z + 5 = 0, \qquad x - y - z = 0.$$

6. Show that the line with equations (20.5) is parallel to the xy-plane, if and only if $|a_1b_2| = 0$; parallel to the yz-plane, or the xz-plane, if and only if $|b_1c_2| = 0$, or $|a_1c_2| = 0$. Under what conditions is it parallel to the y-axis?

7. Using Theorem [20.2], show that the plane $x + 2z = 0$ is parallel to the line

$$x - 2y + 4z + 4 = 0, \qquad x + y + z - 8 = 0$$

(see § 19, Ex. 7).

8. Given two nonparallel lines with direction numbers u_1, v_1, w_1 and u_2, v_2, w_2, show that $|v_1w_2|$, $-|u_1w_2|$, $|u_1v_2|$ are direction numbers of any line perpendicular to the given lines.

9. Show that if two lines with direction numbers u_1, v_1, w_1 and u_2, v_2, w_2 pass through the point (x_1, y_1, z_1), any line through this point and contained in the plane of the two lines has as equations

$$\frac{x - x_1}{t_1u_1 + t_2u_2} = \frac{y - y_1}{t_1v_1 + t_2v_2} = \frac{z - z_1}{t_1w_1 + t_2w_2}$$

for suitable values of t_1 and t_2; and for any values of t_1 and t_2, not both zero, these are equations of a line in this plane and through the point.

10. Show that if the planes (19.1) intersect in a line the equation

$$\frac{a_1x + b_1y + c_1z + d_1}{a_1u + b_1v + c_1w} = \frac{a_2x + b_2y + c_2z + d_2}{a_2u + b_2v + c_2w}$$

is an equation of a plane through this line and parallel to any line with direction numbers u, v, w. Discuss this equation when the planes (19.1) are parallel.

21. Determinants of the Third Order. Three Equations of the First Degree in Three Unknowns

The determination of the common solutions, if any, of three equations

$$\begin{aligned} a_1x + b_1y + c_1z + d_1 &= 0, \\ a_2x + b_2y + c_2z + d_2 &= 0, \\ a_3x + b_3y + c_3z + d_3 &= 0 \end{aligned} \tag{21.1}$$

is facilitated by the use of *determinants of the third order.* Such a determinant is represented by a square array of 9 elements as in the left-hand member of (21.2), and is defined in terms of determinants of the second order as in the right-hand member of (21.2), which is called an *expansion* of the determinant:

$$(21.2)\quad \begin{vmatrix} a_1 & b_1 & c_1 \\ a_2 & b_2 & c_2 \\ a_3 & b_3 & c_3 \end{vmatrix} = a_1 \begin{vmatrix} b_2 & c_2 \\ b_3 & c_3 \end{vmatrix} - a_2 \begin{vmatrix} b_1 & c_1 \\ b_3 & c_3 \end{vmatrix} + a_3 \begin{vmatrix} b_1 & c_1 \\ b_2 & c_2 \end{vmatrix}.$$

We define the *minor* of an element in a determinant of the third order to be the determinant of the second order obtained on removing from the given determinant the row and column in which the element lies; thus in the right-hand member of (21.2) each of the elements of the first column is multiplied by its minor.

It is convenient, as a means of reducing writing, to represent the determinant (21.2) by $|a_1b_2c_3|$, that is, by writing only the elements of the *main diagonal* of the determinant, as was done in § 20 for determinants of the second order. In this notation equation (21.2) is

$$(21.3)\quad |a_1b_2c_3| = a_1|b_2c_3| - a_2|b_1c_3| + a_3|b_1c_2|.$$

It is evident that this abbreviated notation cannot be used when the elements are particular numbers; for, then there would be no means of telling what the elements not on the main diagonal are.

If we pick out the terms in the right-hand member of (21.2) involving b_1, b_2, and b_3 respectively, we see that another way of writing this right-hand member is

$$-b_1(a_2c_3 - a_3c_2) + b_2(a_1c_3 - a_3c_1) - b_3(a_1c_2 - a_2c_1),$$

and consequently another expansion of the determinant is given by

$$(21.4)\quad |a_1b_2c_3| = -b_1|a_2c_3| + b_2|a_1c_3| - b_3|a_1c_2|.$$

If, in like manner, we pick out the terms involving c_1, c_2, c_3, we get the expansion

$$(21.5)\quad |a_1b_2c_3| = c_1|a_2b_3| - c_2|a_1b_3| + c_3|a_1b_2|.$$

On the other hand, the following expansions of the determinant are obtained according as we pick out from the right-hand member of (21.2) the terms involving the elements of the first, second, or third rows:

$$
\begin{aligned}
(21.6) \qquad | a_1b_2c_3 | &= a_1 | b_2c_3 | - b_1 | a_2c_3 | + c_1 | a_2b_3 |; \\
&= - a_2 | b_1c_3 | + b_2 | a_1c_3 | - c_2 | a_1b_3 |; \\
&= a_3 | b_1c_2 | - b_3 | a_1c_2 | + c_3 | a_1b_2 |.
\end{aligned}
$$

When we examine the six expansions of the determinant given by (21.3), (21.4), (21.5), and (21.6), we note that each element multiplied by its minor appears twice: once in the expansions in terms of the elements of a column and their minors, and once in the expansions in terms of the elements of a row and their minors. Also, in both cases the algebraic sign of the term is the same; and we note that if the element is in the pth row and qth column the sign is plus or minus according as $p + q$ is an even or odd number. For example, a_2 is in the second row and first column and, $2 + 1$ being odd, the sign is minus, as is seen to be the case in (21.3) and the second of (21.6). If then we define the *cofactor* of an element in the pth row and qth column to be the minor of the element multiplied by $(-1)^{p+q}$, all six of the above expansions are equivalent to the theorem

[21.1] *A determinant of the third order is equal to the sum of the products of the elements of any column (or row) and their respective cofactors.*

As corollaries of this theorem we have

[21.2] *If all the elements of any column (or row) are zeros, the determinant is equal to zero.*

[21.3] *If all the elements of any column (or row) have a common factor k, the determinant is equal to k times the determinant obtained by removing this factor from all the elements of this column (or row).*

Consider now the determinant

$$
\begin{vmatrix} a_1 & a_2 & a_3 \\ b_1 & b_2 & b_3 \\ c_1 & c_2 & c_3 \end{vmatrix},
$$

that is, the determinant obtained from (21.2) by interchanging its rows and columns, without changing the relative order of the elements in a row or column. When this determinant is expanded in terms of the first row, we have an expression equal to the right-hand member of (21.2), since interchanging rows and columns in a determinant of the second order does not change its value; for example,

$$\begin{vmatrix} b_2 & b_3 \\ c_2 & c_3 \end{vmatrix} = b_2c_3 - b_3c_2 = \begin{vmatrix} b_2 & c_2 \\ b_3 & c_3 \end{vmatrix}.$$

Hence we have

[21.4] *The determinant obtained from a determinant of the third order by interchanging its columns and rows without changing the relative order of the elements in any column or row is equal to the original determinant.*

If this theorem did not hold, we could not use $|a_1b_2c_3|$ to denote the determinant (21.2), since the interchange of columns and rows does not change the main diagonal.

We consider next the result of interchanging two columns (or rows) of a determinant, and begin with the case of two adjacent rows, say, the pth and $(p+1)$th rows, where p is 1 or 2. An element of the qth column and pth row goes into an element of the qth column and $(p+1)$th row. This interchange does not affect the minor of the element; but the cofactor of the element is now the product of the minor by $(-1)^{p+1+q}$, whereas in the original determinant the multiplier is $(-1)^{p+q}$. Consequently the expansion of the new determinant in terms of the elements of the $(p+1)$th row is equal to minus the original determinant. The same result follows when there is an interchange of adjacent columns. The interchange of the first and third rows (or columns) may be effected by three interchanges of adjacent rows (or columns), thus:

$$123 \longrightarrow 132 \longrightarrow 312 \longrightarrow 321.$$

Since three changes of sign result in a change, we have

[21.5] *The determinant obtained by interchanging two rows (or columns) of a determinant of the third order is equal to minus the original determinant.*

As a corollary we have

[21.6] *If all the elements of one row (or column) of a determinant are equal to the corresponding elements of another row (or column), that is, if two rows (or columns) are identical, the determinant is equal to zero.*

In fact, if these rows (or columns) are interchanged, the determinant is evidently the same as before; and in consequence of Theorem [**21.5**] it is equal to that number which is equal to its negative, that is, zero.

We are now able to prove the theorem

[21.7] *The sum of the products of the elements of a row (or column) and the cofactors of the corresponding elements of another row (or column) is equal to zero.*

In fact, such a sum is an expansion of a determinant with two identical rows (or columns), which by Theorem [**21.6**] is equal to zero. This proves the theorem.

With the aid of these properties of determinants we are able to find the common solutions, if any, of equations (21.1). Assuming that they have a common solution, and letting x, y, and z denote the common solution, we multiply equations (21.1) by $|b_2c_3|$, $-|b_1c_3|$, and $|b_1c_2|$ respectively and add. In consequence of (21.3) the coefficient of x is $|a_1b_2c_3|$, and the constant term is $|d_1b_2c_3|$; and from Theorem [**21.7**] it follows that the coefficients of y and z are equal to zero. Hence we have

$$|a_1b_2c_3|\,x + |d_1b_2c_3| = 0. \tag{21.7}$$

Likewise, if we multiply equations (21.1) by $-|a_2c_3|$, $|a_1c_3|$, and $-|a_1c_2|$ respectively and add, we get, in consequence of (21.4) and Theorem [**21.7**],

$$|a_1b_2c_3|\,y + |a_1d_2c_3| = 0. \tag{21.8}$$

Again, if we use the multipliers $|a_2b_3|$, $-|a_1b_3|$, and $|a_1b_2|$, we obtain, in consequence of (21.5) and Theorem [**21.7**],

$$|a_1b_2c_3|\,z + |a_1b_2d_3| = 0. \tag{21.9}$$

We observe that in (21.7), (21.8), and (21.9) x, y, and z have the same coefficient, namely, $|a_1b_2c_3|$, which we call *the determinant of equations* (21.1). Moreover, the second determinants in (21.7), (21.8), and (21.9) are obtained from the determinant of the equations by replacing the a's, b's, and c's respectively by d's having the same subscripts.

Equations (21.7), (21.8), and (21.9) have one and only one solution if $|a_1b_2c_3| \neq 0$. In this case the values of x, y, and z given by these equations satisfy equations (21.1), as the reader can verify (see § 26, Ex. 8). Hence we have

[21.8] *Three equations of the first degree in three unknowns have one and only one common solution if their determinant is not equal to zero.*

The geometric equivalent of this theorem is

[21.9] *Three planes have one and only one point in common if the determinant of their equations is not equal to zero.*

In §§ 22 and 24 we analyze the case when the determinant of equations (21.1) is equal to zero.

Theorem [**21.8**] and the processes leading up to it may be applied to the problem of finding an equation of the plane through three points (x_1, y_1, z_1), (x_2, y_2, z_2), (x_3, y_3, z_3). Thus the coefficients a, b, c, and d in the equation

$$ax + by + cz + d = 0 \tag{21.10}$$

must be such that

$$\begin{aligned} ax_1 + by_1 + cz_1 + d &= 0, \\ ax_2 + by_2 + cz_2 + d &= 0, \\ ax_3 + by_3 + cz_3 + d &= 0. \end{aligned} \tag{21.11}$$

These equations looked upon as equations in a, b, and c have one and only one common solution if the determinant

$$\begin{vmatrix} x_1 & y_1 & z_1 \\ x_2 & y_2 & z_2 \\ x_3 & y_3 & z_3 \end{vmatrix} \tag{21.12}$$

is not equal to zero. If this condition is satisfied, we solve (21.11) for a, b, and c in terms of d and substitute in (21.10) to obtain an equation of the plane (see Ex. 10).

EXERCISES

1. For what value of a is the determinant

$$\begin{vmatrix} 1 & 2 & 0 \\ -1 & 3 & 6 \\ 2 & -1 & a \end{vmatrix}$$

equal to zero? Is it always possible to choose the value of one element in a determinant, all the others being given, so that the determinant shall be equal to a given number?

2. Show that

$$\begin{vmatrix} 1 & 1 & 1 \\ p & q & r \\ p^2 & q^2 & r^2 \end{vmatrix} = (p-q)(q-r)(r-p).$$

3. Show that

$$\begin{vmatrix} a_1 + a_1' & b_1 & c_1 \\ a_2 + a_2' & b_2 & c_2 \\ a_3 + a_3' & b_3 & c_3 \end{vmatrix} = | a_1b_2c_3 | + | a_1'b_2c_3 |.$$

4. Show that the determinant

$$\begin{vmatrix} a_1 & b_1 & c_1 \\ a_2 & b_2 & c_2 \\ t_1a_1 + t_2a_2 & t_1b_1 + t_2b_2 & t_1c_1 + t_2c_2 \end{vmatrix}$$

is equal to zero for any values of t_1 and t_2.

5. Show that whatever be the constants k_1, k_2, k_3

$$\begin{vmatrix} k_1a_1 + k_2b_1 + k_3c_1 & b_1 & c_1 \\ k_1a_2 + k_2b_2 + k_3c_2 & b_2 & c_2 \\ k_1a_3 + k_2b_3 + k_3c_3 & b_3 & c_3 \end{vmatrix} = k_1 | a_1b_2c_3 | + k_2 | b_1b_2c_3 | + k_3 | c_1b_2c_3 |$$

$$= k_1 | a_1b_2c_3 |.$$

6. Show that

$$\begin{vmatrix} | b_1c_2 | & | a_1c_2 | \\ | b_3c_1 | & | a_3c_1 | \end{vmatrix} = c_1 | a_1b_2c_3 |.$$

7. Show that equation (5.3) of a line in the plane through the points (x_1, y_1) and (x_2, y_2) can be written in the form

$$\begin{vmatrix} x & y & 1 \\ x_1 & y_1 & 1 \\ x_2 & y_2 & 1 \end{vmatrix} = 0.$$

8. Without expanding the determinants in the following equations:

$$\begin{vmatrix} x & 2 & -6 \\ y & -4 & 9 \\ 1 & 2 & 3 \end{vmatrix} = 0, \qquad \begin{vmatrix} x & y & 2 \\ -2 & 3 & 1 \\ 2 & 4 & 2 \end{vmatrix} = 0, \qquad \begin{vmatrix} x^2 + y^2 & 2x & y \\ 1 & 2 & -1 \\ 3 & 1 & 2 \end{vmatrix} = 0,$$

show that the first is an equation of the line through the points $(1, -2)$, $(-2, 3)$, and the second an equation of the line through the points $(-4, 6)$, $(2, 4)$. Of what is the third an equation?

9. Show that the planes

$$x + 2y - z + 3 = 0, \qquad 3x - y + 2z + 1 = 0, \qquad 2x - y + z - 2 = 0$$

have one and only one point in common. Find the coordinates of this point by means of (21.7), (21.8), and (21.9), and check the result.

10. Show that when the determinant (21.12) is different from zero for three points (x_1, y_1, z_1), (x_2, y_2, z_2), and (x_3, y_3, z_3), an equation of the plane determined by these points is

$$\begin{vmatrix} 1 & y_1 & z_1 \\ 1 & y_2 & z_2 \\ 1 & y_3 & z_3 \end{vmatrix} x + \begin{vmatrix} x_1 & 1 & z_1 \\ x_2 & 1 & z_2 \\ x_3 & 1 & z_3 \end{vmatrix} y + \begin{vmatrix} x_1 & y_1 & 1 \\ x_2 & y_2 & 1 \\ x_3 & y_3 & 1 \end{vmatrix} z - \begin{vmatrix} x_1 & y_1 & z_1 \\ x_2 & y_2 & z_2 \\ x_3 & y_3 & z_3 \end{vmatrix} = 0.$$

11. Show that if the planes (21.1) have one and only one point in common, the equations

$$\frac{a_1x + b_1y + c_1z + d_1}{a_1u + b_1v + c_1w} = \frac{a_2x + b_2y + c_2z + d_2}{a_2u + b_2v + c_2w} = \frac{a_3x + b_3y + c_3z + d_3}{a_3u + b_3v + c_3w}$$

are equations of a line through this point and with direction numbers u, v, w (see § 20, Ex. 10).

12. Using the notation $\Sigma a_{1i}b_{i1} = a_{11}b_{11} + a_{12}b_{21} + a_{13}b_{31}$ and so on, show that

$$\begin{vmatrix} \Sigma a_{1i}b_{i1} & \Sigma a_{1i}b_{i2} & \Sigma a_{1i}b_{i3} \\ \Sigma a_{2i}b_{i1} & \Sigma a_{2i}b_{i2} & \Sigma a_{2i}b_{i3} \\ \Sigma a_{3i}b_{i1} & \Sigma a_{3i}b_{i2} & \Sigma a_{3i}b_{i3} \end{vmatrix}$$

is equal to the product of the determinants

$$\begin{vmatrix} a_{11} & a_{12} & a_{13} \\ a_{21} & a_{22} & a_{23} \\ a_{31} & a_{32} & a_{33} \end{vmatrix}, \qquad \begin{vmatrix} b_{11} & b_{12} & b_{13} \\ b_{21} & b_{22} & b_{23} \\ b_{31} & b_{32} & b_{33} \end{vmatrix}.$$

13. Show that the result of Ex. 12 may be stated as follows: The product of two determinants of the third order is equal to the determinant of the third order whose element in the ith row and jth column is the sum of the products of corresponding elements of the ith row of the first determinant and the jth column of the second determinant. Discuss the effect of interchanging the two given determinants.

14. Show that if the planes (21.1) have one and only one point in common, the equation

$$t_1(a_1x+b_1y+c_1z+d_1)+t_2(a_2x+b_2y+c_2z+d_2)+t_3(a_3x+b_3y+c_3z+d_3)=0,$$

for any values of t_1, t_2, and t_3, not all zeros, is an equation of a plane through this point, and that an equation of any plane through this point is given by the above equation for suitable values of t_1, t_2, and t_3.

15. Show that if the planes (21.1) have one and only one point in common, an equation of the plane through this point and parallel to the plane

$$\text{(i)} \qquad a_4x + b_4y + c_4z + d_4 = 0$$

is

$$|a_4b_2c_3|(a_1x + b_1y + c_1z + d_1) + |a_1b_4c_3|(a_2x + b_2y + c_2z + d_2) + |a_1b_2c_4|(a_3x + b_3y + c_3z + d_3) = 0.$$

Discuss the case when the plane (i) is parallel to one of the planes (21.1); also when it is parallel to the line of intersection of two of the planes (21.1).

22. Three Homogeneous Equations of the First Degree in Three Unknowns

We consider in this section three homogeneous equations of the first degree

$$\begin{aligned} a_1x + b_1y + c_1z &= 0, \\ a_2x + b_2y + c_2z &= 0, \\ a_3x + b_3y + c_3z &= 0. \end{aligned} \tag{22.1}$$

Since these equations are of the form (21.1) with the d's equal to zero, it follows from Theorem [**21.2**] that the second determinants in (21.7), (21.8), and (21.9) are equal to zero. Consequently $x = y = z = 0$ is the only common solution of equations (22.1) when the determinant of these equations is not equal to zero.

We consider now the case when $|a_1b_2c_3| = 0$. If any two of equations (22.1) are equivalent, that is, if either is a constant multiple of the other, the determinant $|a_1b_2c_3|$ is equal to zero by Theorems [**21.3**] and [**21.6**]. In this case the common solutions of two nonequivalent equations are common solutions of all three equations. Thus, if the third of (22.1) is equivalent to either of the other equations, by Theorem [**20.1**] the common solutions of (22.1) are

$$x = t\,|b_1c_2|, \qquad y = -t\,|a_1c_2|, \qquad z = t\,|a_1b_2| \tag{22.2}$$

for every value of the constant t.

If now no two of equations (22.1) are equivalent, the common solutions of the first two of (22.1) are given by (22.2). When these expressions are substituted in the left-hand member of the third of equations (22.1), we obtain

$$t\,(a_3\,|b_1c_2| - b_3\,|a_1c_2| + c_3\,|a_1b_2|).$$

The expression in parentheses is equal to $|a_1b_2c_3|$ by (21.6). Since by hypothesis this is equal to zero, the above expression is equal to zero for all values of t. Consequently, when $|a_1b_2c_3| = 0$, the expressions (22.2) for every value of t are common solutions of equations (22.1).

In consequence of Theorem [**20.1**] the common solutions of the second and third of equations (22.1) are

$$x = r\,|b_2c_3|, \qquad y = -r\,|a_2c_3|, \qquad z = r\,|a_2b_3| \tag{22.3}$$

for every value of r, and of the first and third of (22.1)

$$x = -s\,|b_1c_3|, \qquad y = s\,|a_1c_3|, \qquad z = -s\,|a_1b_3| \tag{22.4}$$

for every value of s. By the above argument these are common solutions of all three of equations (22.1) when $|a_1b_2c_3| = 0$. The signs in (22.2), (22.3), and (22.4) are chosen so that the determinants in these expressions are the cofactors of the elements in the third, first, and second rows respectively of (21.2.)

As a consequence of (22.2), (22.3), and (22.4) we have that when $|a_1b_2c_3| = 0$ the corresponding determinants in any two of the sets of equations (22.2), (22.3), and (22.4) are proportional (see § 21, Ex. 6). Accordingly we have

[**22.1**] *Three homogeneous equations of the first degree in three unknowns have common solutions other than zero, if and only if the determinant of the equations is equal to zero; for equations* (22.1) *the common solutions are given by*

$$\begin{aligned} x : y : z &= |b_1c_2| : -|a_1c_2| : |a_1b_2| \\ &= |b_2c_3| : -|a_2c_3| : |a_2b_3| \\ &= -|b_1c_3| : |a_1c_3| : -|a_1b_3|. \end{aligned} \tag{22.5}$$

From the foregoing discussion it follows that the three planes (22.1) have the origin as the only common point if the determinant of the equations, that is, $|a_1b_2c_3|$, is different from zero, and that if their determinant is equal to zero the three planes have in common a line through the origin. Equations of such a line are

$$\frac{x}{u} = \frac{y}{v} = \frac{z}{w}.$$

Comparing these equations with (22.2), (22.3), and (22.4), we have as the geometric equivalent of Theorem [**22.1**]

[**22.2**] *Three distinct planes whose equations are* (22.1) *have the origin as the only common point if the determinant of their equations is not equal to zero; and if the determinant is equal to zero the three planes intersect in a line through the origin, having as direction numbers each set of three quantities in equations* (22.5).

As a consequence of Theorems [**22.1**] and [**21.4**] we have

[**22.3**] *When a determinant* $|a_1b_2c_3|$ *is equal to zero, there exist quantities* h_1, h_2, h_3, *not all zero, such that*

$$h_1a_1 + h_2b_1 + h_3c_1 = 0, \quad h_1a_2 + h_2b_2 + h_3c_2 = 0, \quad h_1a_3 + h_2b_3 + h_3c_3 = 0, \tag{22.6}$$

and also quantities k_1, k_2, k_3, *not all zero, such that*

$$k_1a_1 + k_2a_2 + k_3a_3 = 0, \quad k_1b_1 + k_2b_2 + k_3b_3 = 0, \quad k_1c_1 + k_2c_2 + k_3c_3 = 0. \tag{22.7}$$

In fact, the h's are proportional to each set of three quantities in equations (22.5). In like manner the k's are proportional to the cofactors of the elements in each column of $|a_1b_2c_3|$.

Theorem [**22.1**] is very important, and some of its applications may strike the reader as surprising. For example, if we seek an equation of the line in the plane through the points $P_1(x_1, y_1)$ and $P_2(x_2, y_2)$, we have for consideration three equations

$$ax + by + c = 0, \quad ax_1 + by_1 + c = 0, \quad ax_2 + by_2 + c = 0, \tag{22.8}$$

the first of which is an equation of the line whose coefficients are to be determined so that the second and third hold. Applying Theorem [**20.1**] to the last two of equations (22.8), we have

$$a = t\begin{vmatrix} y_1 & 1 \\ y_2 & 1 \end{vmatrix}, \qquad b = t\begin{vmatrix} 1 & x_1 \\ 1 & x_2 \end{vmatrix}, \qquad c = t\begin{vmatrix} x_1 & y_1 \\ x_2 & y_2 \end{vmatrix}.$$

On substituting these values in the first of (22.8) and dividing out the common factor t, we obtain

$$\text{(22.9)} \qquad \begin{vmatrix} y_1 & 1 \\ y_2 & 1 \end{vmatrix} x + \begin{vmatrix} 1 & x_1 \\ 1 & x_2 \end{vmatrix} y + \begin{vmatrix} x_1 & y_1 \\ x_2 & y_2 \end{vmatrix} = 0.$$

However, if we consider (22.8) as three homogeneous equations of the first degree in a, b, and c, the coefficients being $x, y, 1$; $x_1, y_1, 1$; $x_2, y_2, 1$, it follows from Theorem [**22.1**] that, in order that these equations shall have a solution other than $a = b = c = 0$ (in which case we have no equation of the line), we must have

$$\text{(22.10)} \qquad \begin{vmatrix} x & y & 1 \\ x_1 & y_1 & 1 \\ x_2 & y_2 & 1 \end{vmatrix} = 0.$$

This is the same equation as (22.9), as is seen on expanding the determinant (22.10) in terms of the elements of the first row. Consequently we have found an equation of the line by means of Theorem [**22.1**], without finding directly the expressions for a, b, and c (see § 21, Ex. 7).

This is an interesting and subtle process, which may be applied to some of the exercises below, and which is used in the next section. Accordingly it is important that the reader think it through so that he will have confidence in using it.

EXERCISES

1. Find the coefficient a in the first of the equations

$$ax + y + 2z = 0, \qquad 2x - 3y + 4z = 0, \qquad 5x + 2y - 7z = 0$$

so that these equations shall have common solutions other than zeros, and find these solutions.

2. Show that three points (x, y), (x_1, y_1), (x_2, y_2) are vertices of a triangle, if and only if the determinant in (22.10) is not zero. Express the result of § 13, Ex. 26 by means of a determinant.

3. Show that the condition that the determinant in Ex. 2 shall vanish is equivalent to the conditions of Ex. 9 of § 4 that the three points shall be collinear.

4. Show that three distinct nonparallel lines

$$a_1x + b_1y + c_1 = 0, \qquad a_2x + b_2y + c_2 = 0, \qquad a_3x + b_3y + c_3 = 0$$

meet in a point, if and only if

$$\begin{vmatrix} a_1 & b_1 & c_1 \\ a_2 & b_2 & c_2 \\ a_3 & b_3 & c_3 \end{vmatrix} = 0.$$

Compare this result with the discussion of equation (10.4) and with Ex. 4 of § 21.

5. Show that an equation of the plane through the origin and the points (x_1, y_1, z_1) and (x_2, y_2, z_2) is

$$\begin{vmatrix} x & y & z \\ x_1 & y_1 & z_1 \\ x_2 & y_2 & z_2 \end{vmatrix} = 0.$$

6. Show that three points (x_1, y_1, z_1), (x_2, y_2, z_2), (x_3, y_3, z_3) lie in a plane through the origin, if and only if

$$\begin{vmatrix} x_1 & y_1 & z_1 \\ x_2 & y_2 & z_2 \\ x_3 & y_3 & z_3 \end{vmatrix} = 0.$$

7. Show that the condition of Ex. 6 is satisfied when the three points are collinear (see § 15, Ex. 10). Interpret this result and Ex. 6 geometrically.

8. For what values of t do the equations

$$4x + 3y + z = tx, \qquad 3x - 4y + 7z = ty, \qquad x + 7y - 6z = tz$$

admit solutions other than zeros? Find the solutions.

9. Show that if u_1, v_1, w_1; u_2, v_2, w_2; u_3, v_3, w_3 are direction numbers of three lines through a point P, the lines lie in a plane, if and only if the determinant

$$\begin{vmatrix} u_1 & v_1 & w_1 \\ u_2 & v_2 & w_2 \\ u_3 & v_3 & w_3 \end{vmatrix}$$

is equal to zero (see Theorem [**17.7**]).

10. Given three lines through a point P, and not in the same plane, show that direction numbers of any line through P are expressible linearly and homogeneously in terms of the direction numbers of the given three lines.

23. Equations of Planes Determined by Certain Geometric Conditions. Shortest Distance between Two Lines

If we desire to find an equation of the plane determined by three noncollinear points (x_1, y_1, z_1), (x_2, y_2, z_2), (x_3, y_3, z_3), we may substitute these values for x, y, z in the equation

$$ax + by + cz + d = 0, \tag{23.1}$$

solve the resulting equations for a, b, and c in terms of d by the method of § 21, substitute the values of a, b, and c so obtained in (23.1), and get an equation of the plane, as in Ex. 10 of § 21; or we may proceed as follows.

If we substitute x_1, y_1, z_1 for x, y, z in (23.1) and subtract the resulting equation from (23.1), we obtain

$$a(x - x_1) + b(y - y_1) + c(z - z_1) = 0. \tag{23.2}$$

Expressing the conditions that the points (x_2, y_2, z_2) and (x_3, y_3, z_3) are points of the plane (23.2), we have

$$\begin{aligned} a(x_2 - x_1) + b(y_2 - y_1) + c(z_2 - z_1) = 0,\\ a(x_3 - x_1) + b(y_3 - y_1) + c(z_3 - z_1) = 0. \end{aligned} \tag{23.3}$$

Looking upon equations (23.2) and (23.3) as homogeneous equations in a, b, and c, we have in accordance with Theorem [**22.1**] that these equations admit a common solution other than zeros, if and only if their determinant is equal to zero; that is,

$$\begin{vmatrix} x - x_1 & y - y_1 & z - z_1 \\ x_2 - x_1 & y_2 - y_1 & z_2 - z_1 \\ x_3 - x_1 & y_3 - y_1 & z_3 - z_1 \end{vmatrix} = 0. \tag{23.4}$$

When this determinant is expanded in terms of the elements of the first row, it is seen to be an equation of the first degree in x, y, and z, and consequently is an equation of a plane; and it is an equation of the plane determined by the three points, since the determinant is equal to zero when x, y, z are replaced by the coordinates of each of the three points, in accordance with Theorems [**21.2**] and [**21.6**].

Another way of obtaining equation (23.4) is to solve equations (23.3) for a, b, c in accordance with Theorem [**20.1**] and to substitute the result in (23.2), as the reader should verify. But the method we have used leads to the result more immediately.

Equation (23.4) is an equation in x, y, and z unless the minors of $(x - x_1)$, $(y - y_1)$, and $(z - z_1)$ are all equal to zero. If they are equal to zero, the last two rows in (23.4) are proportional, that is,

$$\frac{x_3 - x_1}{x_2 - x_1} = \frac{y_3 - y_1}{y_2 - y_1} = \frac{z_3 - z_1}{z_2 - z_1},$$

which is the condition that the three points shall be collinear. Since there is an endless number of planes through a line, we would not expect equation (23.4) to apply to this case. However, since equations (23.3) are equivalent in this case, if we take any values of a, b, and c satisfying either of equations (23.3) and substitute these values in (23.2), we get an equation of a plane through the three collinear points.

Accordingly we have

[23.1] *Equation* (23.4) *is an equation of the plane determined by three noncollinear points* (x_1, y_1, z_1), (x_2, y_2, z_2), (x_3, y_3, z_3). *When the points are collinear, equation* (23.2) *for values of* a, b, *and* c *satisfying either of equations* (23.3) *is an equation of one of the endless number of planes containing the points.*

We consider next two lines with the equations

$$\frac{x - x_1}{u_1} = \frac{y - y_1}{v_1} = \frac{z - z_1}{w_1}, \tag{23.5}$$

$$\frac{x - x_2}{u_2} = \frac{y - y_2}{v_2} = \frac{z - z_2}{w_2}, \tag{23.6}$$

and seek an equation of the plane through the first line and parallel to the second. Since the plane passes through the point (x_1, y_1, z_1), an equation of the plane is of the form

$$a(x - x_1) + b(y - y_1) + c(z - z_1) = 0. \tag{23.7}$$

If the line (23.5) is to be in this plane, and the line (23.6) is to be parallel to the plane, any normal to the plane must be perpendicular to both lines; consequently a, b, and c, direction numbers of such a normal, must satisfy the conditions

$$\begin{aligned} au_1 + bv_1 + cw_1 &= 0, \\ au_2 + bv_2 + cw_2 &= 0, \end{aligned} \tag{23.8}$$

by Theorems [16.8] and [17.5]. Looking upon these equations and (23.7) as homogeneous equations in a, b, and c, we have that these equations admit solutions not all zeros, if and only if

$$\begin{vmatrix} x - x_1 & y - y_1 & z - z_1 \\ u_1 & v_1 & w_1 \\ u_2 & v_2 & w_2 \end{vmatrix} = 0, \tag{23.9}$$

which is the equation sought.

When two lines do not intersect and are not parallel, by constructing a plane through each line parallel to the other we have the lines lying in two parallel planes, like the ceiling and floor of a room.

The shortest distance between the lines is the length of the segment they determine on their common perpendicular; that is, in the above analogy it is the length of the normal to the floor and ceiling which meets the two lines. This length is the distance between the two planes. If then we wish to find the shortest distance between the lines (23.5) and (23.6), and we observe that (23.9) is an equation of the plane through the line (23.5) parallel to the line (23.6), it is clear that this shortest distance is the distance of any point on the line (23.6) from the plane (23.9), and in particular the distance of the point (x_2, y_2, z_2) from this plane. In accordance with Theorem [18.1] this value of the directed distance is obtained by substituting x_2, y_2, z_2 for x, y, z in the left-hand member of (23.9) and dividing by the square root, with appropriate sign, of the sum of the squares of the coefficients of x, y, and z in equation (23.9). Hence we have

[23.2] *The directed shortest distance from the line* (23.5) *to the line* (23.6) *is given by*

$$\frac{1}{D}\begin{vmatrix} x_2 - x_1 & y_2 - y_1 & z_2 - z_1 \\ u_1 & v_1 & w_1 \\ u_2 & v_2 & w_2 \end{vmatrix}, \tag{23.10}$$

where $D = e\sqrt{|v_1w_2|^2 + |w_1u_2|^2 + |u_1v_2|^2}$,

e being $+1$ *or* -1 *according as* $|u_1v_2|$ *is positive or negative, and so on, as in* [18.1].

EXERCISES

1. Find an equation of the plane through the points $(h, 1, 1)$, $(k, 2, -1)$, and $(1, 3, -3)$. For what values of h and k is there more than one plane through these points? Find an equation of one of the planes.

2. Under what condition is equation (23.9) satisfied by any values of x, y, and z? What does this mean geometrically?

3. Show that an equation of a plane containing the line (23.5) and perpendicular to the plane

$$\text{(i)} \qquad ax + by + cz + d = 0$$

is

$$\begin{vmatrix} x - x_1 & y - y_1 & z - z_1 \\ u_1 & v_1 & w_1 \\ a & b & c \end{vmatrix} = 0.$$

Discuss the case when the line (23.5) is normal to the plane (i).

4. Find the shortest distance between the lines

$$\frac{x-1}{2} = \frac{y+1}{1} = \frac{z-2}{-2}, \qquad \frac{x+2}{3} = \frac{y}{1} = \frac{z-3}{5}.$$

5. Show that an equation of the plane through the point (x_1, y_1, z_1) and normal to the line

$$a_1x + b_1y + c_1z + d_1 = 0, \qquad a_2x + b_2y + c_2z + d_2 = 0$$

is

$$\begin{vmatrix} x - x_1 & y - y_1 & z - z_1 \\ a_1 & b_1 & c_1 \\ a_2 & b_2 & c_2 \end{vmatrix} = 0.$$

6. Show that the lines (23.5) and (23.6) lie in the same plane, if and only if

$$\begin{vmatrix} x_1 - x_2 & y_1 - y_2 & z_1 - z_2 \\ u_1 & v_1 & w_1 \\ u_2 & v_2 & w_2 \end{vmatrix} = 0.$$

7. Show that an equation of the plane through the point (x_1, y_1, z_1) and parallel to lines with direction numbers u_1, v_1, w_1 and u_2, v_2, w_2 is

$$\begin{vmatrix} x - x_1 & y - y_1 & z - z_1 \\ u_1 & v_1 & w_1 \\ u_2 & v_2 & w_2 \end{vmatrix} = 0.$$

8. Show that for the points $P_1(x_1, y_1, z_1)$, $P_2(x_2, y_2, z_2)$, and so on

$$\begin{vmatrix} x - x_1 & x_2 - x_3 & x_4 - x_5 \\ y - y_1 & y_2 - y_3 & y_4 - y_5 \\ z - z_1 & z_2 - z_3 & z_4 - z_5 \end{vmatrix} = 0$$

is an equation of a plane through P_1 parallel to P_2P_3 and to P_4P_5.

9. Find the coordinates of the point or points common to the two planes $2x+3y+z+3=0$, $x-2y-3z-2=0$, and each of the planes

(a) $x+y-2z+3=0$; (c) $-2x+4y+6z+4=0$;
(b) $3x+y-2z+1=0$; (d) $x+5y+4z+6=0$.

What is the geometric relation between each set of three planes?

10. When u_1, v_1, w_1 and u_2, v_2, w_2 in (23.5) and (23.6), are direction cosines, how may the expression (23.10) be written in consequence of (15.10)?

11. Show by means of Ex. 3 that an equation of the plane containing the line (23.5) and perpendicular to the plane (23.9) is

$$\begin{vmatrix} x-x_1 & y-y_1 & z-z_1 \\ u_1 & v_1 & w_1 \\ |v_1w_2| & |w_1u_2| & |u_1v_2| \end{vmatrix} = 0.$$

Show that this equation, and the one obtained from it on replacing x_1, y_1, z_1; u_1, v_1, w_1 by x_2, y_2, z_2; u_2, v_2, w_2 respectively in the first two rows, are equations of the common perpendicular to the lines (23.5) and (23.6).

12. What equations replace (23.9) and the equation of Ex. 11 when the lines (23.5) and (23.6) are parallel?

24. The Configurations of Three Planes

In this section we return to the consideration of three equations

$$(24.1) \qquad \begin{aligned} a_1x+b_1y+c_1z+d_1&=0,\\ a_2x+b_2y+c_2z+d_2&=0,\\ a_3x+b_3y+c_3z+d_3&=0 \end{aligned}$$

and seek the conditions upon the coefficients in these equations corresponding to the various types of configurations of three planes. In § 21 it was shown that a common solution or solutions, if any, of these equations satisfy the equations

$$(24.2) \qquad \begin{aligned} |a_1b_2c_3|\,x+|d_1b_2c_3|&=0,\\ |a_1b_2c_3|\,y+|a_1d_2c_3|&=0,\\ |a_1b_2c_3|\,z+|a_1b_2d_3|&=0; \end{aligned}$$

and from these equations followed Theorem [21.9], that three planes with equations (24.1) have one and only one point in common if the determinant of these equations, namely, $|a_1b_2c_3|$, is not equal to zero.

It is evident geometrically that there are the following configurations formed by three planes, other than that of planes having one and only one point in common: (1) two of the planes are parallel, and the third intersects them in parallel lines; (2) the three planes are parallel; (3) the three planes have a line in common, or intersect in three parallel lines forming a triangular prism. In accordance with Theorem [21.9] these cases must correspond to the various ways in which the equation

$$\begin{vmatrix} a_1 & b_1 & c_1 \\ a_2 & b_2 & c_2 \\ a_3 & b_3 & c_3 \end{vmatrix} = 0 \tag{24.3}$$

is satisfied. We shall consider these various ways successively, and observe first of all that if (24.3) is satisfied there is no common solution of equations (24.1) unless all the determinants

$$|d_1b_2c_3|, \quad |a_1d_2c_3|, \quad |a_1b_2d_3| \tag{24.4}$$

in (24.2) are equal to zero.

Case 1. The condition (24.3) is satisfied when any two rows are proportional. If the first two rows are proportional, that is, if $|a_1b_2| = |b_1c_2| = |a_1c_2| = 0$, by Theorem [**19.4**] the first two planes are parallel or coincident according as d_1/d_2 is not or is equal to the ratio of the other coefficients; that is, as one can show, when the minors of the elements a_3, b_3, and c_3 in the determinants (24.4) are not or are equal to zero; although there are six of these minors, only three are distinct. If the third row is not proportional to the other two, the third plane meets the other two planes in two parallel lines or in two coincident lines according as the first two planes are parallel or coincident. In these cases equations (24.1) have no common solution or an endless number of common solutions respectively.

Case 2. If all three rows in (24.3) are proportional, that is, if the minors of all the elements in (24.3) are equal to zero, the three planes are parallel, or two are coincident and parallel to the third, or all three are coincident. These cases are distinguished from one another by the values of the ratios of the

d's. Thus, if the minors of a_3, b_3, c_3 in the determinants (24.4) are not all zero, the first two planes (24.1) are parallel (and not coincident), and similarly for other pairs of planes. Only in case all three planes are coincident do equations (24.1) have common solutions.

Case 3. We consider finally the case when no two rows are proportional. From Theorem [**22.3**] we have equations (22.7), in which all the k's are different from zero, since otherwise one finds that two rows are proportional. If we solve equations (22.7) for a_3, b_3, and c_3, and put $-k_1/k_3 = t_1$, $-k_2/k_3 = t_2$, we have

$$a_3 = t_1 a_1 + t_2 a_2, \qquad b_3 = t_1 b_1 + t_2 b_2, \qquad c_3 = t_1 c_1 + t_2 c_2. \tag{24.5}$$

When these expressions are substituted in the determinants (24.4), the latter reduce respectively to

$$k\,|\,b_1 c_2\,|, \qquad -k\,|\,a_1 c_2\,|, \qquad k\,|\,a_1 b_2\,|, \tag{24.6}$$

where the common factor k is given by

$$k = d_3 - t_1 d_1 - t_2 d_2. \tag{24.7}$$

We consider first the case when the three quantities (24.6) are equal to zero, that is, when the determinants (24.4) are equal to zero. Since by hypothesis the first two rows of (24.3) are not proportional, we must have $k = 0$; that is,

$$d_3 = t_1 d_1 + t_2 d_2. \tag{24.8}$$

From this result and (24.5) we have

$$\begin{aligned} a_3 x + b_3 y + c_3 z + d_3 = t_1(a_1 x + b_1 y + c_1 z + d_1) \\ + t_2(a_2 x + b_2 y + c_2 z + d_2). \end{aligned}$$

Hence by Theorem [**19.6**] the three planes (24.1) have a line in common.

When $k \neq 0$, that is, when the determinants (24.4) are not zero, from (24.5), and (24.7) solved for d_3, we have

$$\begin{aligned} a_3 x + b_3 y + c_3 z + d_3 = t_1(a_1 x + b_1 y + c_1 z + d_1) \\ + t_2(a_2 x + b_2 y + c_2 z + d_2) + k, \end{aligned} \tag{24.9}$$

from which it is seen that the three equations (24.1) do not have a common solution; for, if there were a common solution, the expression on the left in (24.9) and the expressions in parentheses would be equal to zero for the values of x, y, and z of this solution, which is impossible for $k \neq 0$. Geometrically this means that the three planes intersect in three parallel lines; that is, they form a triangular prism (see Ex. 2).

The results of the foregoing discussion may be stated as follows:

[24.1] *When the determinant* (24.3) *of equations* (24.1) *of three planes is equal to zero, it follows that:*

(1) If the minors of the elements of any row are equal to zero, the planes corresponding to the other two rows are coincident or parallel according as the minors of the elements of the corresponding rows in the determinants (24.4) *are equal to zero or not; if only two of the planes are parallel and all three are distinct, the third plane intersects the other two in parallel lines.*

(2) If the minors of all the elements of all the rows are equal to zero, the three planes are parallel, or two are coincident and parallel to the third, or all three are coincident.

(3) If the minors of all the elements in the determinant (24.3) *are not all zero, the planes meet one another in one line or in three parallel lines according as all the determinants* (24.4) *are equal to zero or not.*

We have also the algebraic theorem

[24.2] *Three equations* (24.1), *not all of which are equivalent and whose determinant is equal to zero, have an endless number of common solutions or none according as all three determinants* (24.4) *are equal to zero or not; in the former case either two of the equations are equivalent or any one is a linear combination of the other two, that is, a sum of constant multiples of the other two.*

Theorems [**21.8**] and [**24.2**] constitute a complete statement about the common solutions of three nonhomogeneous equations of the first degree in three unknowns, and Theorems [**21.9**] and [**24.1**] give a geometric picture of this algebraic problem.

EXERCISES

1. Discuss Ex. 9 of § 23 in the light of the above analysis.

2. Show that for any nonzero values of t_1, t_2, and t_3 the equation

$$t_1(a_1x + b_1y + c_1z + d_1) + t_2(a_2x + b_2y + c_2z + d_2) + t_3 = 0$$

is an equation of a plane parallel to the line

$$a_1x + b_1y + c_1z + d_1 = 0, \qquad a_2x + b_2y + c_2z + d_2 = 0.$$

(See § 19, Ex. 7.)

3. Using Theorem [**20.2**], show that when (24.1) are equations of three distinct planes and equation (24.3) is satisfied, the cofactors of any row in the determinant of the equations are direction numbers of the line, or lines, of intersection of the planes when such cofactors are not all zero.

4. For what values of a, b, and c in the equations

$$ax + 2y + 3z - 1 = 0, \quad 3x + by + z + 2 = 0, \quad 11x + 8y + cz - 3 = 0$$

are the planes with these equations mutually perpendicular? For what values do the planes meet in one line? For what values do they meet in three parallel lines?

25. Miscellaneous Exercises. The Sphere

1. Find an equation of the set of all planes through the point (2, – 1, 3) parallel to the line

$$3x - 2y + z - 1 = 0, \quad x + y - 2z + 2 = 0.$$

2. Find an equation of the plane through the origin normal to the line $\quad 3x - y + 4z + 5 = 0, \quad x + y + z = 0.$

3. Prove that the line

$$x + 2y - z + 3 = 0, \quad 3x - y + 2z + 1 = 0$$

meets the line

$$2x - 2y + 3z - 2 = 0, \quad x - y - z + 3 = 0.$$

4. Find the point where the line $\dfrac{x-3}{2} = \dfrac{y-4}{3} = \dfrac{z-5}{6}$ meets the plane $x + y + z = 0$. How far is this point from the point (3, 4, 5)?

5. Show that

$$\begin{vmatrix} 1 & 1 & 1 \\ p & q & r \\ p^3 & q^3 & r^3 \end{vmatrix} = (p-q)(q-r)(r-p)(p+q+r).$$

6. What is the locus of a point the sum of whose distances from any number n of planes is a constant?

7. Show that the projection (see § 14) of a line segment of length l upon a line L is equal to $l \cos \theta$, where θ is the angle of the line L and the segment; and that the projection upon a plane is $l \sin \phi$, where ϕ is the angle of the segment and a normal to the plane.

8. Find the projection of the line segment between the points $(2, -5, 1)$ and $(4, -1, 5)$ upon the line $\frac{x+1}{2} = \frac{y-2}{-1} = \frac{z-1}{2}$; also upon the plane $2x - y + 2z = 0$.

9. Show that if g, h, k are the intercepts of a plane on the x-, y-, and z-axes respectively, and p is the distance of the origin from the plane,

$$\frac{1}{g^2} + \frac{1}{h^2} + \frac{1}{k^2} = \frac{1}{p^2}.$$

10. Given a fixed point P on a line in space through the origin and equally inclined to the three coordinate axes, show that for every plane through P meeting the three axes the sum of the reciprocals of the intercepts has the same value.

11. Find the condition that the three lines

$$\frac{x}{a_1} = \frac{y}{b_1} = \frac{z}{c_1}, \quad \frac{x}{a_2} = \frac{y}{b_2} = \frac{z}{c_2}, \quad \frac{x}{a_3} = \frac{y}{b_3} = \frac{z}{c_3}$$

shall lie in a plane.

12. Show that the bisectors of the angles between perpendicular lines through the origin and with direction cosines λ_1, μ_1, ν_1 and λ_2, μ_2, ν_2 respectively have equations

$$\frac{x}{\lambda_1 \pm \lambda_2} = \frac{y}{\mu_1 \pm \mu_2} = \frac{z}{\nu_1 \pm \nu_2}.$$

13. Find an equation of the sphere of radius r and center at the point (x_0, y_0, z_0).

14. Show that

(i) $$x^2 + y^2 + z^2 + 2fx + 2gy + 2hz + e = 0$$

is an equation of a sphere. What are the center and radius? (See § 12.)

15. Show that the points common to two intersecting spheres lie in a plane which is perpendicular to the line through the centers of the spheres. This plane is called the *radical plane.* Discuss this question when the spheres do not intersect (see § 12).

16. Show that the square of the length of any tangent to a sphere with an equation (i) of Ex. 14 from a point (x_1, y_1, z_1) outside the sphere is equal to $x_1^2 + y_1^2 + z_1^2 + 2fx_1 + 2gy_1 + 2hz_1 + e$ (see § 12).

17. Find an equation of the plane normal to the radius of the sphere $x^2 + y^2 + z^2 - 2x + 4y + 2z + 2 = 0$ through the point $(1, -2, 1)$ of the sphere. This plane is called the *tangent plane* to the sphere at this point.

18. Find an equation of the sphere which passes through the origin and the points $(1, -2, 3)$, $(2, 0, -1)$, $(4, 4, 0)$; of the sphere when the last point is replaced by $(3, 2, -5)$.

19. Let $S_1 = 0$ and $S_2 = 0$ be equations of two spheres in the form (i) of Ex. 14. Discuss the equation $t_1S_1 + t_2S_2 = 0$ when t_1 and t_2 take all values, not both zero, and in particular the case $t_1 = -t_2$.

20. Find an equation of the sphere inscribed in the tetrahedron whose faces are the coordinate planes and the plane $x - 2y + 2z = 4$.

21. Find the locus of a point the square of whose distance from the origin is equal to its distance from the plane $x - 2y + 2z = 0$.

22. Find the locus of a point the sum of the squares of whose distances from the points $(1, 0, -1)$ and $(2, 1, -3)$ is 10.

23. What is the locus of a point the sum of the squares of whose distances from any number n of points is a constant?

24. Find an equation of the right circular cylinder of radius r and with the line $x - x_1 = 0$, $y - y_1 = 0$ for axis.

25. Show that $ax^2 + by^2 + cz^2 = 0$, where a, b, and c do not all have the same sign, is an equation of a cone with vertex at the origin, by showing that if $P_1(x_1, y_1, z_1)$ is a point of the locus, so also is every point on the line joining the origin and P_1. Could the locus be one or more planes?

26. Find the locus of a point whose distance from the z-axis is equal to its distance from the xy-plane. For what part of the locus are the directed distances equal?

27. Find the locus of a point whose distance from the origin is twice its distance from the xy-plane.

28. What is the character of the locus with the equation $f(x, z) = 0$, where $f(x, z)$ denotes an expression in x and z; of the locus $f(y) = 0$?

26. Determinants of Any Order

In § 21 we defined determinants of the third order in terms of the elements of the first column and their minors, these being determinants of the second order, and derived various theorems concerning determinants of the third order. In this section we define determinants of the fourth and higher orders, and show that the theorems of § 21 apply equally well to these determinants.

We begin with a determinant of the fourth order, represented by a square array of 16 elements and defined as follows in terms of determinants of the third order:

$$(26.1)\quad \begin{vmatrix} a_1 & b_1 & c_1 & d_1 \\ a_2 & b_2 & c_2 & d_2 \\ a_3 & b_3 & c_3 & d_3 \\ a_4 & b_4 & c_4 & d_4 \end{vmatrix} = a_1 \begin{vmatrix} b_2 & c_2 & d_2 \\ b_3 & c_3 & d_3 \\ b_4 & c_4 & d_4 \end{vmatrix} - a_2 \begin{vmatrix} b_1 & c_1 & d_1 \\ b_3 & c_3 & d_3 \\ b_4 & c_4 & d_4 \end{vmatrix} + a_3 \begin{vmatrix} b_1 & c_1 & d_1 \\ b_2 & c_2 & d_2 \\ b_4 & c_4 & d_4 \end{vmatrix} - a_4 \begin{vmatrix} b_1 & c_1 & d_1 \\ b_2 & c_2 & d_2 \\ b_3 & c_3 & d_3 \end{vmatrix}$$

If we expand the determinants in the right-hand member of (26.1) in terms of the elements of the first columns, and collect the terms in b_1, b_2, b_3, and b_4, we have, using the main diagonal to represent a determinant,

$$\begin{aligned}(26.2)\quad |a_1b_2c_3d_4| = &- b_1(a_2 |c_3d_4| - a_3 |c_2d_4| + a_4 |c_2d_3|) \\ &+ b_2(a_1 |c_3d_4| - a_3 |c_1d_4| + a_4 |c_1d_3|) \\ &- b_3(a_1 |c_2d_4| - a_2 |c_1d_4| + a_4 |c_1d_2|) \\ &+ b_4(a_1 |c_2d_3| - a_2 |c_1d_3| + a_3 |c_1d_2|) \\ = &- b_1 |a_2c_3d_4| + b_2 |a_1c_3d_4| - b_3 |a_1c_2d_4| \\ &+ b_4 |a_1c_2d_3|\end{aligned}$$

This is an expansion of the determinant in terms of the elements of the second column and their minors.

If we expand the determinants of the third order in (26.1) in terms of the elements of the second columns, and again in terms of the elements of the third columns, and proceed as above, we obtain

$$
\begin{aligned}
(26.3)\quad |a_1b_2c_3d_4| &= c_1\,|a_2b_3d_4| - c_2\,|a_1b_3d_4| + c_3\,|a_1b_2d_4| \\
&\qquad - c_4\,|a_1b_2d_3|; \\
&= -d_1\,|a_2b_3c_4| + d_2\,|a_1b_3c_4| - d_3\,|a_1b_2c_4| \\
&\qquad + d_4\,|a_1b_2c_3|.
\end{aligned}
$$

In the foregoing expansions it is seen that the element of the pth row and qth column, for any values of p and q from 1 to 4, is multiplied by $(-1)^{p+q}$ times the *minor* of the element, which by definition is the determinant of the third order obtained on removing from the determinant (26.1) the row and column in which the element lies. For example, b_3 is in the third row and second column, in which case $(-1)^{p+q} = (-1)^{3+2} = -1$, and we see that this checks with (26.2).

As in § 21, we define the *cofactor* of the element in the pth row and qth column to be $(-1)^{p+q}$ times the minor of the element. Accordingly, although the determinant was defined as the sum of the products of the elements of the first column and their respective cofactors, it is shown by (26.2) and (26.3) that the determinant is equal to the sum of the products of the elements of any column and their respective cofactors.

Since, as we have just seen, the terms involving any element consist of the products of this element and its cofactor, it follows also, as in § 21, that the determinant is equal to the sum of the products of the elements of any row and their respective cofactors (see Theorem [**21.1**]).

Just as determinants of the third and fourth orders have been defined to be the sum of the products of the elements of the first column and their respective cofactors, generalizing the notation and terminology, we define a *determinant of any order to be the sum of the products of the elements of the first column and their respective cofactors*. Just as we have shown that it follows from the definition of determinants of the third and fourth orders that the following theorem holds, so by proceeding step by step with determinants of the fifth order, and so on, we can establish the following theorem for determinants of any order:

[26.1] *A determinant of any order is equal to the sum of the products of the elements of any row (or column) and their respective cofactors.*

We consider in connection with the determinant (26.1) the determinant obtained from (26.1) by interchanging rows and columns without changing the relative order of the rows and columns, that is, the determinant

$$(26.4) \qquad \begin{vmatrix} a_1 & a_2 & a_3 & a_4 \\ b_1 & b_2 & b_3 & b_4 \\ c_1 & c_2 & c_3 & c_4 \\ d_1 & d_2 & d_3 & d_4 \end{vmatrix}.$$

When this determinant is expanded in terms of the elements of the first row, we have in place of the right-hand member of (26.1) an expression obtained from the latter when the rows and columns in each of the four determinants of the third order are interchanged. But by Theorem [**21.4**] these determinants of the third order are equal respectively to the determinants from which they were obtained by the interchange. Hence the determinant (26.4) is equal to the determinant (26.1). Proceeding step by step, we can show that this result holds for a determinant of any order. Hence we have

[26.2] *The determinant obtained from a given determinant by interchanging its rows and columns without changing the relative position of the elements in the rows and columns is equal to the given determinant.*

Consider now the effect of interchanging two adjacent columns of a determinant. An element in the new determinant is in the same row as originally, but the number of its column is one less, or one greater, than in the given determinant. Consequently, if $(-1)^{p+q}$ is the multiplier of its minor yielding its cofactor in the original determinant, the multiplier in the new determinant is $(-1)^{p+q-1}$ or $(-1)^{p+q+1}$. Since $(-1)^{p+q-1} = -(-1)^{p+q} = (-1)^{p+q+1}$, it follows that in either case the sign of the cofactor is changed. If then we expand the two determinants in terms of the same elements, we have that the new determinant is -1 times the original determinant.

Consider next the determinant resulting from a given determinant by the interchange of any two nonadjacent columns, say the rth and the sth, where $s > r$. The elements of the sth column can be brought into the rth column by $s - r$ interchanges of adjacent columns. This leaves the elements of the original rth column in the $(r+1)$th column, and then, by $s - r - 1$ interchanges of adjacent columns, these elements can be brought into the sth column. Since this interchange of the rth and sth columns can be accomplished by $2(s-r) - 1$ interchanges of adjacent columns, and since each such interchange introduces -1 as a multiplier, the result is to multiply the original determinant by -1 raised to the odd power $2(s-r)-1$; that is, to multiply the original determinant by -1.

Since similar results are obtained when two rows are interchanged, we have the theorem

[26.3] *The determinant obtained by interchanging two rows (or columns) of a determinant is equal to minus the original determinant.*

As a corollary we have

[26.4] *When two rows (or columns) of a determinant are identical, the determinant is equal to zero.*

For, on interchanging the two rows (or columns) the sign is changed, but we have the same determinant over again; and zero is the only number which is equal to its negative.

From Theorems [**26.1**] and [**26.4**] we have

[26.5] *The sum of the products of the elements of any row (or column) of a determinant and the cofactors of the corresponding elements of another row (or column) is equal to zero.*

For, the sum of such products is an expansion of a determinant with two identical rows (or columns).

Since each term in the expansion of a determinant contains one element, and only one, from each column and each row, it follows that

[26.6] *The multiplication of each element of a row (or column) of a determinant by a constant k is equivalent to the multiplication of the determinant by k.*

Accordingly, if all the elements of a row (or column) have the same factor k, the determinant is equal to k times the determinant which results on removing this factor from each element of this row (or column).

As a consequence of Theorems [**26.4**] and [**26.6**] we have

[26.7] *When the elements of two rows (or columns) of a determinant are proportional, the determinant is equal to zero.*

In consequence of Theorem [**26.1**] we have (see § 21, Ex. 3)

[26.8] *If each element of any row (or column) of a determinant is expressed as the sum of two quantities, the determinant may be written as the sum of two determinants.*

If one wishes to write a determinant of any order n, it is convenient to designate an element by one letter having two subscripts, for example, by a_{ij}, where i denotes the row and j the column in which the element occurs. In this notation from Theorem [**26.8**] we have

$$\begin{vmatrix} a_{11} + ka_{12} & a_{12} & a_{13} \cdots a_{1n} \\ a_{21} + ka_{22} & a_{22} & a_{23} \cdots a_{2n} \\ \cdot\ \cdot\ \cdot & \cdot\ \cdot & \cdot\ \cdot\ \cdot\ \cdot \\ \cdot\ \cdot\ \cdot & \cdot\ \cdot & \cdot\ \cdot\ \cdot\ \cdot \\ a_{n1} + ka_{n2} & a_{n2} & a_{n3} \cdots a_{nn} \end{vmatrix} = |\, a_{11}a_{22} \cdots a_{nn} \,| + k \,|\, a_{12}a_{22}a_{33} \cdots a_{nn} \,|.$$

Since the second of these determinants, having two columns identical, is equal to zero, we have by similar procedure applied to any two rows (or columns) the following theorem:

[26.9] *If to the elements of any row (or column) of a determinant there be added equal multiples of the corresponding elements of any other row (or column), the determinant is unaltered.*

This result is frequently used to replace a determinant by an equivalent one with one or more zero elements, an operation which reduces the calculation involved in evaluating the determinant. For example, consider the following determinant and the equivalent one obtained by multiplying the second row by 2, subtracting the result from the first and fourth rows, and adding the result to the third row:

$$\begin{vmatrix} 2 & 4 & -2 & 3 \\ 1 & -2 & 1 & 0 \\ -2 & 0 & -1 & 3 \\ 2 & 3 & -2 & 3 \end{vmatrix} = \begin{vmatrix} 0 & 8 & -4 & 3 \\ 1 & -2 & 1 & 0 \\ 0 & -4 & 1 & 3 \\ 0 & 7 & -4 & 3 \end{vmatrix} = -1 \begin{vmatrix} 8 & -4 & 3 \\ -4 & 1 & 3 \\ 7 & -4 & 3 \end{vmatrix}$$

$$= -3 \begin{vmatrix} 8 & -4 & 1 \\ -4 & 1 & 1 \\ 7 & -4 & 1 \end{vmatrix}.$$

If in the last third-order determinant we subtract the first row from the second and third rows, the result is

$$-3 \begin{vmatrix} 8 & -4 & 1 \\ -12 & 5 & 0 \\ -1 & 0 & 0 \end{vmatrix} = -3 \begin{vmatrix} -12 & 5 \\ -1 & 0 \end{vmatrix} = -15.$$

EXERCISES

1. Evaluate the following determinants:

$$\begin{vmatrix} 2 & 1 & 4 & -3 \\ -3 & 2 & 2 & 4 \\ 1 & 1 & 3 & -2 \\ -2 & 0 & 1 & 3 \end{vmatrix}, \quad \begin{vmatrix} 1 & 1 & 2 & 3 \\ 2 & 0 & -4 & -1 \\ 3 & -2 & -5 & 3 \\ -1 & -1 & 2 & -3 \end{vmatrix}, \quad \begin{vmatrix} 1 & 2 & 3 & 1 & -2 \\ 2 & 1 & 3 & 2 & 1 \\ 0 & 1 & 2 & 2 & 1 \\ 0 & -1 & -1 & -2 & 1 \\ 0 & 2 & 3 & -1 & -1 \end{vmatrix}.$$

2. For what value of a is the determinant

$$\begin{vmatrix} a & -1 & 4 & 9 \\ 7 & 5 & -2 & -3 \\ -3 & 2 & 4 & -1 \\ 4 & 7 & 2 & 4 \end{vmatrix}$$

equal to zero?

3. Show with the aid of Theorem [**26.9**] that each of the following determinants is equal to zero:

$$\begin{vmatrix} a & b & c & d \\ b & a & c & d \\ b & a & d & c \\ a & b & d & c \end{vmatrix}, \qquad \begin{vmatrix} 1 & a & b & c+d \\ 1 & b & c & a+d \\ 1 & c & d & a+b \\ 1 & d & a & b+c \end{vmatrix}.$$

4. Show that

$$\begin{vmatrix} 1 & 1 & 1 & 1 \\ a & b & c & d \\ a^2 & b^2 & c^2 & d^2 \\ a^3 & b^3 & c^3 & d^3 \end{vmatrix} = (a-b)(a-c)(a-d)(b-c)(b-d)(c-d).$$

5. Show by means of Theorems [**26.6**] and [**26.9**] that if $d_4 \neq 0$ the determinant (26.1) is equal to

$$\frac{1}{(d_4)^2} \begin{vmatrix} |a_1d_4| & |b_1d_4| & |c_1d_4| \\ |a_2d_4| & |b_2d_4| & |c_2d_4| \\ |a_3d_4| & |b_3d_4| & |c_3d_4| \end{vmatrix}.$$

6. Find the ratio of the determinants

$$\begin{vmatrix} a_1fg & b_1g & c_1gk \\ a_2fh & b_2h & c_2hk \\ a_3fk & b_3k & c_3k^2 \end{vmatrix}, \qquad \begin{vmatrix} c_1fg & c_2fk & c_3fh \\ a_1g & a_2k & a_3h \\ b_1g & b_2k & b_3h \end{vmatrix}.$$

7. Show that a determinant is equal to an algebraic sum of all terms each of which consists of the product of one element, and only one, from each row and each column, and that every such product is a term of the sum.

8. Show that, if the determinant $|a_1b_2c_3|$ of equations (21.1) is not equal to zero, and we multiply any one of these equations by $|a_1b_2c_3|$ and substitute in this equation the values of x, y, and z from (21.7), (21.8), and (21.9), the result is expressible as a determinant of the fourth order with two identical rows. Does this prove that these values of x, y, and z constitute the solution of equations (21.1)?

9. Show that

$$\begin{vmatrix} 1 & -x_1 & -x_2 & -x_3 \\ x_1 & a_{11} & a_{12} & a_{13} \\ x_2 & a_{21} & a_{22} & a_{23} \\ x_3 & a_{31} & a_{32} & a_{33} \end{vmatrix} = \sum_{i,j}^{1,2,3} A_{ij}x_ix_j + A,$$

where $A = |a_{11}a_{22}a_{33}| \equiv |a_{ij}|$, and A_{ij} is the cofactor of a_{ij} in A.

10. Show that the determinant in Ex. 9 is equal to

$$\begin{vmatrix} a_{11} + x_1^2 & a_{12} + x_1x_2 & a_{13} + x_1x_3 \\ a_{21} + x_2x_1 & a_{22} + x_2^2 & a_{23} + x_2x_3 \\ a_{31} + x_3x_1 & a_{32} + x_3x_2 & a_{33} + x_3^2 \end{vmatrix} \equiv | a_{ij} + x_ix_j |,$$

and that consequently the latter determinant is equal to

$$\sum_{i,j}^{1,2,3} A_{ij}x_ix_j + A.$$

11. Show that if in Ex. 9 $a_{ij} = a_{ji}$ for all values of i and j, then $A_{ij} = A_{ji}$.

12. Show that the result of Ex. 10 holds for $i, j = 1, \cdots, n$, where n is any positive integer.

13. Show that the rule for the multiplication of two determinants of the third order stated in Ex. 13 of § 21 applies to two determinants of any order, both determinants being of the same order.

27. Solution of Equations of the First Degree in Any Number of Unknowns. Space of Four Dimensions

Consider the four equations

$$\begin{aligned} a_1x + b_1y + c_1z + d_1w + e_1 &= 0, \\ a_2x + b_2y + c_2z + d_2w + e_2 &= 0, \\ a_3x + b_3y + c_3z + d_3w + e_3 &= 0, \\ a_4x + b_4y + c_4z + d_4w + e_4 &= 0, \end{aligned} \tag{27.1}$$

with the understanding that not all the coefficients of x, y, z, and w in any equation are equal to zero. These equations may or may not have a common solution. We assume that they have at least one common solution and that x, y, z, w in equations (27.1) denotes a common solution We multiply equations (27.1) by the cofactors of a_1, a_2, a_3, and a_4 in the determinant $| a_1b_2c_3d_4 |$ respectively and add the resulting equations. In this sum the coefficient of x is $| a_1b_2c_3d_4 |$, *the determinant of equations* (27.1), in consequence of Theorem [26.1], and the coefficients of y, z, and w are equal to zero, in consequence of Theorem [**26.5**]. Hence we have the equation

$$|\,a_1b_2c_3d_4\,|\,x + |\,e_1b_2c_3d_4\,| = 0, \tag{27.2}$$

where $|\,e_1b_2c_3d_4\,|$ is the determinant obtained from $|\,a_1b_2c_3d_4\,|$ on replacing each a by an e with the same subscript.

If in like manner we multiply equations (27.1) respectively by the cofactors of b_1, b_2, b_3, and b_4 in the determinant $|\,a_1b_2c_3d_4\,|$ and add the results, we obtain the equation

$$|\,a_1b_2c_3d_4\,|\,y + |\,a_1e_2c_3d_4\,| = 0. \tag{27.3}$$

Similarly we have

$$\begin{aligned} |\,a_1b_2c_3d_4\,|\,z + |\,a_1b_2e_3d_4\,| &= 0, \\ |\,a_1b_2c_3d_4\,|\,w + |\,a_1b_2c_3e_4\,| &= 0. \end{aligned} \tag{27.4}$$

If the determinant $|\,a_1b_2c_3d_4\,|$ is not equal to zero, these equations have one and only one common solution. Moreover, this solution is a solution of equations (27.1). In fact, if we multiply the left-hand member of the first of (27.1) by $|\,a_1b_2c_3d_4\,|$, and substitute from (27.2), (27.3), and (27.4), we have

$$\begin{aligned} &- a_1\,|\,e_1b_2c_3d_4\,| - b_1\,|\,a_1e_2c_3d_4\,| - c_1\,|\,a_1b_2e_3d_4\,| \\ &\qquad - d_1\,|\,a_1b_2c_3e_4\,| + e_1\,|\,a_1b_2c_3d_4\,|\,. \end{aligned}$$

When this is rewritten in the form

$$\begin{aligned} &a_1\,|\,b_1c_2d_3e_4\,| - b_1\,|\,a_1c_2d_3e_4\,| + c_1\,|\,a_1b_2d_3e_4\,| \\ &\qquad - d_1\,|\,a_1b_2c_3e_4\,| + e_1\,|\,a_1b_2c_3d_4\,|, \end{aligned}$$

it is seen to be the expansion in terms of the elements of the first row of the determinant,

$$\begin{vmatrix} a_1 & b_1 & c_1 & d_1 & e_1 \\ a_1 & b_1 & c_1 & d_1 & e_1 \\ a_2 & b_2 & c_2 & d_2 & e_2 \\ a_3 & b_3 & c_3 & d_3 & e_3 \\ a_4 & b_4 & c_4 & d_4 & e_4 \end{vmatrix}$$

with the first two rows identical, and consequently is equal to zero. Since similar results follow for the other equations (27.1), we have that equations (27.1) have one and only one common solution when their determinant $|\,a_1b_2c_3d_4\,|$ is not equal to zero. Evidently this process may be applied to any number n of equations of the first degree in n unknowns. As a consequence we have the theorem

[27.1] *n equations of the first degree in n unknowns have one and only one common solution when the determinant of the equations is not equal to zero.*

When the determinant $| a_1b_2c_3d_4 |$ is equal to zero, there is no common solution of equations (27.1) if any one of the second determinants in equations (27.2), (27.3), and (27.4) is not equal to zero. A similar statement applies to n equations in n unknowns whose determinant is equal to zero.

We consider next the case when the equations are homogeneous in the unknowns, that is, when all the e's in equations (27.1) are equal to zero. We write them thus:

$$a_ix + b_iy + c_iz + d_iw = 0 \qquad (i = 1, 2, 3, 4). \tag{27.5}$$

In this case all the second determinants in (27.2), (27.3), and (27.4) are zero; and, in consequence of Theorem **[27.1]**, $x = y = z = w = 0$ is the only common solution of equations (27.5) when the determinant of these equations, that is, $| a_1b_2c_3d_4 |$, is not equal to zero.

In order to consider the case when the determinant $| a_1b_2c_3d_4|$ is equal to zero, we denote by A_1 the cofactor of a_1 in this determinant, by C_3 the cofactor of c_3, etc. The equation

$$a_iA_4 + b_iB_4 + c_iC_4 + d_iD_4 = 0 \tag{27.6}$$

for $i = 4$ states that the determinant $| a_1b_2c_3d_4 |$ is equal to zero, in consequence of Theorem **[26.1]**. The three equations (27.6) for $i = 1$, 2, and 3 are identities in consequence of Theorem **[26.5]**. When we compare equations (27.5) and (27.6) as i takes the values 1, 2, 3, and 4, we see that a common solution of equations (27.5) for which $| a_1b_2c_3d_4 | = 0$ is given by

$$x = tA_4, \qquad y = tB_4, \qquad z = tC_4, \qquad w = tD_4, \tag{27.7}$$

for each value of t. This result would seem to indicate that zeros are the only common solution of equations (27.5) when

$A_4 = B_4 = C_4 = D_4 = 0$; but we shall show that this is not the correct conclusion.

From equations (27.7) we obtain

$$D_4x = A_4w, \qquad D_4y = B_4w, \qquad D_4z = C_4w. \tag{27.8}$$

Since by definition A_4, B_4, C_4, and D_4 are the cofactors of the elements of the last row in the determinant (26.1), they are

$$\begin{aligned} A_4 &= -\,|\,b_1c_2d_3\,|, \qquad & C_4 &= -\,|\,a_1b_2d_3\,|, \\ B_4 &= |\,a_1c_2d_3\,|, \qquad & D_4 &= |\,a_1b_2c_3\,|. \end{aligned} \tag{27.9}$$

Accordingly equations (27.8) are those which one obtains when, using the method of § 21, one solves for x, y, z in terms of w the three equations (27.5) as i takes the values 1, 2, 3. From this point of view it follows that when $A_4 = B_4 = C_4 = D_4 = 0$, either one of the three equations under consideration is a constant multiple of one of the others, or any one of the equations is equal to the sum of certain constant multiples of the other two (see Ex. 7). In either case a common solution of two of the equations (not any two in the first case) is a solution of the third equation, and the three equations are *not independent.* Accordingly we have established the theorem

[27.2] *Three independent homogeneous equations*

$$\begin{aligned} a_1x + b_1y + c_1z + d_1w &= 0, \\ a_2x + b_2y + c_2z + d_2w &= 0, \\ a_3x + b_3y + c_3z + d_3w &= 0 \end{aligned} \tag{27.10}$$

admit an endless number of common solutions given by

$$x : y : z : w = -\,|\,b_1c_2d_3\,| : |\,a_1c_2d_3\,| : -\,|\,a_1b_2d_3\,| : |\,a_1b_2c_3\,|; \tag{27.11}$$

and these are solutions also of the equation

$$a_4x + b_4y + c_4z + d_4w = 0,$$

if and only if the determinant of the four equations is equal to zero.

Observe that this theorem is a generalization of Theorem [**20.1**].

Returning to the consideration of equations (27.5), we remark that if we replace A_4, B_4, C_4, D_4 in (27.6) by any one of the three sets A_j, B_j, C_j, D_j respectively, as j takes on the values 1, 2, and 3, the resulting equations are satisfied, one of them because $|a_1b_2c_3d_4| = 0$, and the other three in consequence of Theorem [**26.5**]. Hence not only does (27.7) give a solution of (27.5), but also

$$x = t_jA_j, \qquad y = t_jB_j, \qquad z = t_jC_j, \qquad w = t_jD_j$$

for $j = 1, 2, 3$. If for any j we have $A_j = B_j = C_j = D_j = 0$, it follows from the above discussion that the three equations (27.5), as i takes on values different from the particular value of j, are not independent.

The preceding arguments apply to a set of n homogeneous equations of the first degree in n unknowns and we have

[**27.3**] *When the determinant of n homogeneous equations of the first degree in n unknowns is equal to zero and the cofactors of the elements of any row are not all equal to zero, these cofactors multiplied by an arbitrary constant constitute a common solution of the equations, in addition to the solution consisting only of zeros.*

As a consequence of Theorem [**27.3**] we have the following generalization of Theorem [**22.3**]:

[**27.4**] *When a determinant of the nth order is equal to zero, there exist numbers $h_1, \cdots, h_n$, not all zero, such that the sum of the products of $h_1, \cdots, h_n$ and the corresponding elements in the 1st to nth columns of each and every row is equal to zero; and likewise numbers $k_1, \cdots, k_n$, not all zero, such that the sum of the products of $k_1, \cdots, k_n$ and the corresponding elements of the 1st to nth rows of each and every column is equal to zero.*

We return to the consideration of equations (27.2), (27.3), and (27.4) when the determinant $|a_1b_2c_3d_4|$ is equal to zero. In accordance with Theorem [**27.4**] there exist numbers k_1, k_2, k_3, k_4, not all equal to zero, such that when equations (27.1) are multiplied by k_1, k_2, k_3, k_4 respectively and added, the coefficients of x, y, z, and w in the sum are zero; consequently the assumption that there is a common solution is valid only in case

$$k_1e_1 + k_2e_2 + k_3e_3 + k_4e_4 = 0.$$

This and $| a_1b_2c_3d_4 | = 0$ are the conditions that the four equations (27.1) are not independent; in other words, common solutions of three of the equations are solutions of the fourth. But three equations in four unknowns admit an endless number of solutions of at least one degree of arbitrariness; for, one at least of the unknowns may be chosen arbitrarily, and then the others are fixed by the equations.

Since all of the foregoing discussion applies equally to n equations of the first degree in n unknowns, we have

[27.5] *n equations of the first degree in n unknowns have one and only one common solution, if and only if the determinant of the equations is not equal to zero. If the determinant is equal to zero, there are no common solutions, or an endless number of one or more degrees of arbitrariness. If in the latter case the equations are homogeneous, there is an endless number of solutions.*

In the preceding section and the present one we have shown how the theory of determinants and of first-degree equations which was developed in Chapter 1 in the study of the geometry of the plane and in Chapter 2 in the study of the geometry of space may be generalized algebraically to determinants of any order and to linear equations in any number n of unknowns. In order to speak of these generalizations geometrically, we introduce the concept of spaces of four, five, and any number n dimensions. Although we may not be able to visualize the geometry of such spaces, we may speak about it and deal with it.

Since in two-dimensional space an equation of the first degree defines a line, such an equation is sometimes called *linear* and a line a *linear entity*. These terms are used in three dimensions to designate an equation of the first degree in three unknowns and the plane represented by such an equation respectively. In spaces of four, five, and higher dimensions it is customary to call an equation of the first degree in the corresponding number of unknowns a *linear equation*, and the locus defined by this equation a *linear entity*. Thus in two-dimensional space a linear equation defines a *linear entity* of one lower dimension — a line. In three-dimensional space a linear equation defines a two-dimensional linear entity — the plane; and two independent linear equations define a line. Likewise in space of n dimensions one linear equation defines a linear entity of $n-1$ dimensions, in the sense that each solution of the equation gives the coordinates of a point in this entity, and $n-1$ of the unknowns may be chosen arbitrarily, and then the other is determined. We say that the space of n dimensions *envelops* such an $n-1$ linear entity, and that this entity is *embedded* in the space; for example, the xy-plane is a linear entity of two dimensions embedded in space of coordinates x, y, z. Similarly, in *n-space*, that is, space of n dimensions, two independent linear equations define a linear entity of $n-2$ dimensions, and so on; and, in particular, $n-1$ independent linear equations define a line. In fact, if we denote by $x^1, \cdots, x^n$ the coordinates of a representative point of n-dimensional space, and by $x_1^1, x_1^2, \cdots, x_1^n$ and $x_2^1, x_2^2, \cdots, x_2^n$ the coordinates of two particular points, equations of the line through these points are

$$\frac{x^1 - x_1^1}{x_2^1 - x_1^1} = \frac{x^2 - x_1^2}{x_2^2 - x_1^2} = \cdots = \frac{x^n - x_1^n}{x_2^n - x_1^n},$$

which are a generalization of equations (5.2) and (16.1).

Just as the concepts of direction cosines and direction numbers of a line introduced in Chapter 1 have been generalized to space of three dimensions in this chapter, so they may be generalized to space of any number of dimensions, and therefrom the measure of angle between lines. Thus the analogue of Theorems [**6.9**] and [**17.5**] is that the coefficients $a_1, \cdots, a_n$ in the equation

$$a_1x^1 + a_2x^2 + \cdots + a_nx^n + b = 0$$

are direction numbers of the normals to the $n-1$ linear entity defined by the above equation, there being one of these normals at each point of this entity.

We speak also of a generalized sphere of radius r and center at the point $x_0^1, \cdots, x_0^n$, its equation being

$$(x^1 - x_0^1)^2 + (x^2 - x_0^2)^2 + \cdots + (x^n - x_0^n)^2 = r^2,$$

a generalization of (12.1) (see § 25, Ex. 13).

These are only suggestions of the manner in which the geometric concepts of three-dimensional space may be generalized to a space of n dimensions. The subject is a fascinating one, which the reader may pursue further either by himself (see Exs. 9–15) or in consultation with books and articles dealing with the subject (see the reference list which follows the exercises).

EXERCISES

1. Solve by means of determinants the equations

$$2x - 4y + 3z + 4w = -3, \quad 5x + 8y + 9z + 3w = 9,$$
$$3x - 2y + 6z + 5w = -1, \quad x - 10y - 3z - 7w = 2.$$

2. Show that an equation of the plane determined by three noncollinear points (x_1, y_1, z_1), (x_2, y_2, z_2), (x_3, y_3, z_3) is

$$\begin{vmatrix} x & y & z & 1 \\ x_1 & y_1 & z_1 & 1 \\ x_2 & y_2 & z_2 & 1 \\ x_3 & y_3 & z_3 & 1 \end{vmatrix} = 0.$$

Show that this equation is the same as the one in Ex. 10 of § 21. Discuss this equation when the three points are collinear (see § 15, Ex. 10).

3. Show that a necessary condition that the four planes whose equations are $a_ix + b_iy + c_iz + d_i = 0$, as i takes the values 1, 2, 3, 4, shall have at least one point in common is $| a_1b_2c_3d_4 | = 0$. In what manner can the above condition be satisfied without the four planes' having a point in common? Under what condition have the four planes a line in common?

4. Given the tetrahedron whose vertices are $O(0, 0, 0)$, $A(a, 0, 0)$, $B(0, b, 0)$, and $C(0, 0, c)$, show that the six planes each passing through an edge of the tetrahedron and bisecting the opposite edge meet in a point. Is this result true when the axes are oblique, that is, when they are not mutually perpendicular? Is it true for any tetrahedron?

5. Show that for the tetrahedron of Ex. 4 the six planes each bisecting one edge and perpendicular to this edge meet in a point.

6. Show that in the plane an equation of the circle through the three noncollinear points (x_1, y_1), (x_2, y_2), (x_3, y_3) is

$$\begin{vmatrix} x^2 + y^2 & x & y & 1 \\ x_1^2 + y_1^2 & x_1 & y_1 & 1 \\ x_2^2 + y_2^2 & x_2 & y_2 & 1 \\ x_3^2 + y_3^2 & x_3 & y_3 & 1 \end{vmatrix} = 0.$$

Discuss this equation when the three points are collinear. What is the corresponding equation of a sphere through four noncoplanar points?

7. Show by means of Theorem [**22.3**] that if $A_4 = B_4 = C_4 = D_4 = 0$ (see (27.9)) there is a linear relation between the three equations (27.5) for $i = 1, 2, 3$.

8. Show that when all the determinants in equations (27.2), (27.3), and (27.4) are equal to zero there is a linear relation between two or more of the equations (27.1).

9. Show that the coordinates of any point of a line in n-dimensional space are expressible linearly and homogeneously in terms of the coordinates of two fixed points of the line (see Theorem [**5.4**] and equations (16.9)).

10. In four-dimensional space of coordinates x, y, z, and w the entity defined by $ax + by + cz + dw + e = 0$ is called a *hyperplane*. Show that it possesses the property used by Euclid to characterize a plane in 3-space (see § 17).

11. Show that in four-dimensional space two equations

$$a_1x + b_1y + c_1z + d_1w + e_1 = 0, \qquad a_2x + b_2y + c_2z + d_2w + e_2 = 0,$$

such that the coefficients of the unknowns are not proportional, are equations of a plane (see page 74).

12. Show that in four-dimensional space two planes ordinarily meet in one and only one point. Discuss the exceptional cases.

13. Where are the points in space of four dimensions for which $0 < x < 1$, $y = z = w = 0$; $0 < x < 1$, $0 < y < 1$, $z = w = 0$; $0 < x < 1$, $0 < y < 1$, $0 < z < 1$, $w = 0$; x, y, z, w are all greater than zero and less than 1?

14. Generalize to n-space Theorems [**15.1**], [**15.2**], [**16.3**], [**17.5**], [**17.6**], and [**18.1**].

15. How is a linear entity of r ($< n$) dimensions defined algebraically in n-space?

REFERENCES

JORDAN, C. "Essai sur la géometrie à n dimensions," *Bulletin de la Société Mathématique de France*, Vol. 3 (1875).

MANNING, H. P. Geometry of Four Dimensions. The Macmillan Company, 1914.

SOMMERVILLE, D. M. Y. An Introduction to the Geometry of n Dimensions. Methuen, 1929.

VEBLEN, O., and WHITEHEAD, J. H. C. The Foundations of Differential Geometry, Chapters 1 and 2. Cambridge University Press, 1932.

CAIRNS, S. S. "The Direction Cosines of a p-Space in Euclidean n-Space," *American Mathematical Monthly*, Vol. 39 (1932), pp. 518–523.

CHAPTER 3

Transformations of Coordinates

28. Transformations of Rectangular Coordinates

In defining rectangular coordinates in the plane in § 2 we chose a point of the plane for origin, one line through it for the x-axis, and the line perpendicular to the latter and through the origin for the y-axis, and defined the x- and y-coordinates of any point of the plane so as to be the directed distances of the point from the y-axis and x-axis respectively. Since the origin and axes may be chosen arbitrarily, there is no such thing as *the* coordinate axes for a plane, in the sense that a plane has a definite set of axes predetermined. If then we set up two different sets of axes, it is evident that a given point of the plane will have different coordinates with respect to the two sets of axes. Since an equation of a locus has for its solutions the coordinates of every point of the locus, we should expect the equations of a given locus for two different sets of axes to be different. Just what the difference is will be revealed if we know the relation between the coordinates of each point with respect to the two coordinate systems. Knowing this relation and because the choice of a coordinate system is arbitrary, we are able at times to choose a coordinate system with respect to which an equation of a locus is in simple form (see § 32). It is this relation which we shall obtain in what follows.

We consider first the case when the two coordinate systems have different origins but the x- and y-axes of the two systems are parallel, as shown in Fig. 20. O is the origin of the system of coordinates x and y, and O' of the system of coordinates x' and y'; furthermore the coordinates of O' in the xy-system are x_0 and y_0. Then for a representative point P of coordinates x, y and x', y' in the respective systems we have

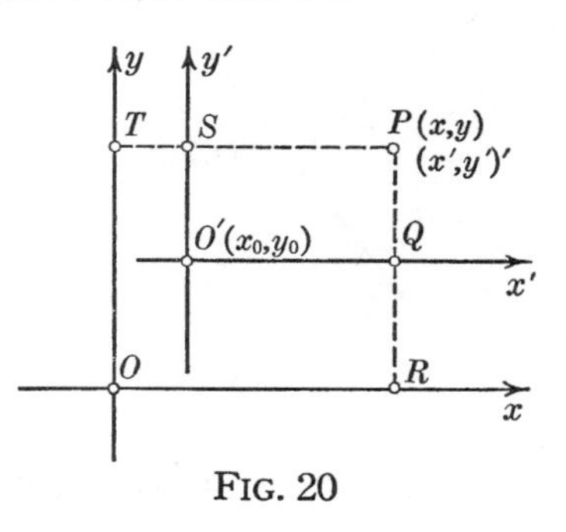

FIG. 20

$$x = TS + SP = x_0 + x', \qquad y = RQ + QP = y_0 + y'.$$

We use parentheses with a prime, as in Fig. 20, to denote a point in the $x'y'$-system.

An equation of any line or curve referred to the xy-system is transformed into an equation of the line or curve referred to the $x'y'$-system by the substitution

$$x = x' + x_0, \qquad y = y' + y_0. \tag{28.1}$$

Although these equations have been derived for the case when P is in the first quadrant of each system, the reader can easily verify that they are valid for any position of P.

When equations (28.1) are solved for x' and y', we obtain

$$x' = x - x_0, \qquad y' = y - y_0, \tag{28.2}$$

which could have been obtained directly from Fig. 20 on noting that the coordinates of O relative to the $x'y'$-system are $-x_0, -y_0$. Also equations of the x-axis and y-axis with respect to the $x'y'$-system are $y' + y_0 = 0$ and $x' + x_0 = 0$ respectively.

The transformation of coordinates (28.1), the inverse of which is given by (28.2), is sometimes called a *translation* or *parallel displacement* of the axes.

We consider next the general situation when the two sets of axes are not parallel, as shown in Figs. 21 and 22. In each

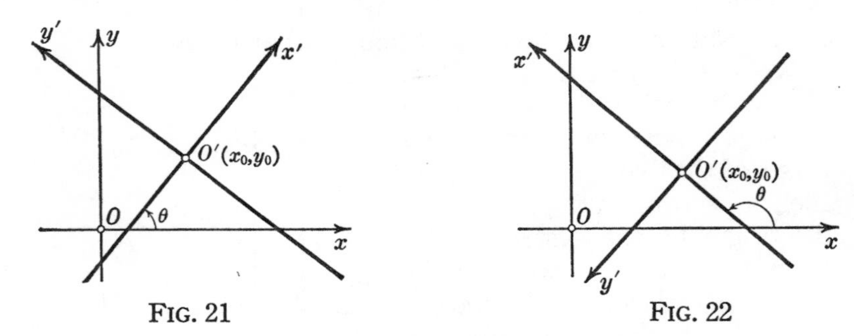

FIG. 21 FIG. 22

case the direction cosines of the x'-axis relative to the xy-system are $\cos\theta$ and $\sin\theta$, and an equation of the x'-axis is

$$\frac{x - x_0}{\cos\theta} = \frac{y - y_0}{\sin\theta},$$

or in other form

$$(y - y_0)\cos\theta - (x - x_0)\sin\theta = 0. \tag{28.3}$$

In each case by Theorem [6.11] an equation of the y'-axis, which is the perpendicular to the x'-axis at O', is

$$(y - y_0) \sin \theta + (x - x_0) \cos \theta = 0. \tag{28.4}$$

Since $\cos \theta$, the coefficient of y in equation (28.3), is positive for Fig. 21, and since $\sin^2 \theta + \cos^2 \theta = 1$, it follows from Theorem [8.1] that for any values of x and y the expression $(y - y_0) \cos \theta - (x - x_0) \sin \theta$ is the directed distance of the point $P(x, y)$ from the x'-axis, and that it is a positive or negative number according as P is above or below this line. By definition this is the coordinate y' of P in the $x'y'$-system. Hence we have the relation

$$y' = (y - y_0) \cos \theta - (x - x_0) \sin \theta. \tag{28.5}$$

Moreover, since $\sin \theta$, the coefficient of y in (28.4), is positive for Fig. 21, by the same kind of reasoning applied to (28.4) we have

$$x' = (y - y_0) \sin \theta + (x - x_0) \cos \theta. \tag{28.6}$$

Consider now Fig. 22. Since $\sin \theta$ is positive also in this case, the right-hand member of (28.6) for any values of x and y is the directed distance of the point $P(x, y)$ from the y'-axis, that is, the line (28.4), the distance being positive or negative according as P lies above or below the y'-axis. Since the positive direction of the x'-axis is above the y'-axis, equation (28.6) holds for this case also. Since $\cos \theta$ is negative, when the coordinates x and y of a point P above the line which is the x'-axis are substituted in the left-hand member of equation (28.3), the resulting number is minus the directed distance from the line to the point, as follows from the considerations of § 8. In the $x'y'$-system P lies below the x'-axis, and consequently the y' of P is negative, and its absolute value is equal to the absolute value of the left-hand member of (28.3). Hence equation (28.5) holds also for Fig. 22.

Similarly it can be shown that (28.5) and (28.6) hold when the $x'y'$-system is in such position that θ lies between 180° and 360°.

Solving equations (28.5) and (28.6) for $x - x_0$ and $y - y_0$, we write the result in the form

$$\begin{aligned} x &= x' \cos \theta - y' \sin \theta + x_0, \\ y &= x' \sin \theta + y' \cos \theta + y_0. \end{aligned} \tag{28.7}$$

Hence an equation of any line or curve referred to the xy-system is transformed into an equation in the $x'y'$-system when the expressions (28.7) are substituted for x and y in the given equation (see § 40). Equations (28.1) are the special case of (28.7) when $\theta = 0$.

When, in particular, the origin O' coincides with O, equations (28.7) reduce to

$$(28.8) \quad x = x' \cos\theta - y' \sin\theta, \quad y = x' \sin\theta + y' \cos\theta.$$

This transformation is sometimes referred to as a *rotation* of the coordinate axes. In this case equations (28.6) and (28.5) are

$$(28.9) \quad x' = x \cos\theta + y \sin\theta, \quad y' = -x \sin\theta + y \cos\theta.$$

These equations may also be obtained from (28.8) by interchanging x and y with x' and y' respectively and replacing θ by $-\theta$, which is what we should expect from the fact that the x-axis makes the angle $-\theta$ with the x'-axis.

Equations (28.8) and (28.9) of rotation of the axes may be written in the condensed form

	x'	y'
x	$\cos\theta$	$-\sin\theta$
y	$\sin\theta$	$\cos\theta$

with the understanding that each equation (28.8) is obtained by equating x (or y) to the sum of the products obtained by multiplying the element in the square which is in the same row as x (or y) by the coordinate x' or y' directly above the element. Equations (28.9) are obtained by taking the sum of the elements in the same column as x' or y' after multiplying each of them by the coordinate on its left.

Consider, for example, the transformation of coordinates when the x'-axis is the line $x + 2y - 4 = 0$ and the origin O' is the point (2, 1) on this line. Then an equation of the y'-axis is $2x - y - 3 = 0$, it being the perpendicular to the x'-axis through the point (2, 1). The slope of the x'-axis is negative; and if we take as the positive sense of the x'-axis that of the line $x + 2y - 4 = 0$, we are dealing with the type represented in Fig. 22 in which the positive direction of the y'-axis is downward. Accordingly we have the second of the following equations, the first following from the fact that the positive direction of the x'-axis is upward:

$$x' = -\frac{2x - y - 3}{\sqrt{5}}, \qquad y' = -\frac{x + 2y - 4}{\sqrt{5}}.$$

EXERCISES

1. Find an equation of the line $2x - 3y - 5 = 0$ with reference to the $x'y'$-system with origin at the point $(1, -1)$ and with axes parallel to the x- and y-axes.

2. Show that it follows from (28.8) that $x^2 + y^2 = x'^2 + y'^2$, and explain why the quantity $x^2 + y^2$ should be *invariant* under a rotation of the coordinate axes.

3. Find an equation of the curve $x^2 + y^2 + 2x - 4y = 0$ when referred to a coordinate system with axes parallel to the x- and y-axes and with origin at the point $(-1, 2)$.

4. Find an equation of the curve $y^2 + 2y - 8x - 15 = 0$ when referred to a coordinate system with origin at the point $(-2, -1)$ and with axes parallel to the x- and y-axes.

5. Find the transformation of coordinates to axes with origin at $(1, -1)$ and with the line $4x + 3y - 1 = 0$ for the new y-axis, the positive sense along the latter being that of the line $4x + 3y - 1 = 0$.

6. Find the transformation of coordinates of oblique axes into axes parallel to them (see § 11).

7. Two systems of rectangular coordinates, (x, y) and $(x', y')'$, are related so that the points $(0, 0)$, $(1, 0)$, $(1, 1)$ in the xy-system are the points $(1, -1)'$, $(1, 0)'$, $(0, 0)'$ respectively in the $x'y'$-system. Draw a diagram showing the relative positions of the two sets of axes, and find the equations of the corresponding transformation.

8. Show that equations (28.5), (28.6), and (28.7) for θ a negative angle between $-180°$ and $0°$ are the same as those for an appropriate θ between $180°$ and $360°$.

9. Determine the translation of the axes such that the equation $2x^2 - 3y^2 - 4x - 12y = 0$ is transformed into one in which there are no terms of the first degree in x' and y'.

10. Into what equation is the equation $9x^2+2\sqrt{3}\,xy+11y^2-4=0$ transformed when the axes are rotated through 60°?

11. Justify the following statement without carrying through the transformation of coordinates involved: In terms of coordinates x' and y', referring to a coordinate system with origin at the center of the first of the circles (12.11), and with the line of centers of the two circles for x'-axis, equations of the circles are

$$x'^2+y'^2+k_1'=0, \quad x'^2+y'^2+2f_2'x'+k_2'=0;$$

and from this result it follows that the centers of the circles (12.12) lie on a line.

12. Show that equations (28.5) and (28.6) may be interpreted as the result of a translation of the xy-system to the point (x_0, y_0) for new origin, and then a rotation of axes through the angle θ. What are the equations if first there is a rotation of the axes through the angle θ and then a translation to (x_0, y_0) as new origin? Compare the resulting equations with (28.7).

13. Find the transformation of coordinates from a rectangular xy-system to oblique axes with the equations $ax+by=0$, $cx+dy=0$. What is the area of the triangle whose vertices are $(1, 0)'$, $(0, 0)'$, $(0, 1)'$ in the new system?

29. Polar Coordinates in the Plane

Rectangular and oblique coordinate systems are not the only kinds of coordinate systems which may be employed in the treatment of geometric loci. Another system frequently used is one in which the coordinates of a point P are its distance r from a fixed point O called the *origin*, or *pole*, and the angle θ which the line segment, or *vector*, OP makes with a fixed vector OM, the *polar axis* of the system, the angle θ being measured from the axis in the counterclockwise direction. r and θ so defined, and called the *radius vector* and *vectorial angle* respectively, are *polar coordinates* of P, and (r, θ) denotes the point with these coordinates. The reader in drawing graphs of algebraic equations in x and y may have used graph paper with two sets of parallel lines, each set perpendicular to the other, which form an array of small squares.

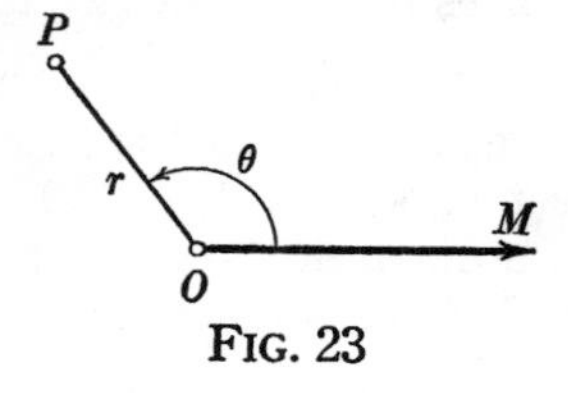

FIG. 23

There exists also graph paper for polar coordinates ruled with a set of concentric circles and with lines radiating from the common center of the circles.

We observe that the point P may be defined also by r and $\theta + n\,360°$ for any positive integer n. In this sense a point may have many sets of polar coordinates, all referred to the same pole and axis. Also negative values of θ may be used to assign polar coordinates to a point, a negative angle being described from the axis OM in the clockwise direction. Thus far we have understood r to be a positive number, but it is advisable to give a meaning to polar coordinates when r is negative. By definition, if r is negative, we lay off from O the length $|r|$ not on the vector making the angle θ with OM but on the vector making the angle $\theta + 180°$. With this understanding $-2, \theta$ and $2, \theta + 180°$ are polar coordinates of the same point. The necessity for a convention concerning negative values of r arises when one seeks the graph of an equation in polar coordinates, that is, the locus of all points with solutions of the equation as polar coordinates.

Consider, for example, the equation

$$r = a \cos \theta \qquad (a > 0). \tag{29.1}$$

For a value of θ, say θ_1, in the first quadrant r is positive, and r for $\theta_1 + 180°$ has the same numerical value but is negative; consequently, for two such values of θ one obtains the same point. Similar reasoning applies when θ is in the second quadrant. Another way of stating this result is that as θ takes the values from 0° to 360° the curve is described twice. Clearly there is no necessity in this case of taking values of θ greater than 180°, because of the periodic property of $\cos \theta$. This does not apply to such an equation as $r = \cos \frac{\theta}{3}$ (see Ex. 6).

Consider next the equation

$$r^2 = 4 \cos \theta. \tag{29.2}$$

In order that r may be real, $\cos \theta$ cannot be negative; consequently admissible values of θ are from $-90°$ to $+90°$. For each such θ we have two values of r, namely, $\pm 2\sqrt{\cos \theta}$. Consequently, on the line making such an angle θ with the axis OM

there are two points of the curve on opposite sides of O and equidistant from it (see Fig. 24). The curve is symmetric with respect to the origin, to the polar axis, and to the line $\theta = 90°$. In Fig. 24, on each line there is noted the angle which it makes with the axis, expressed in degrees and in radians.

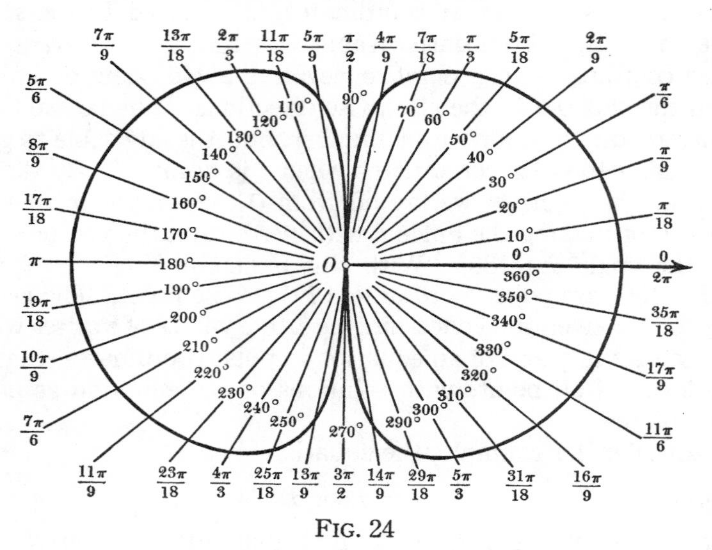

FIG. 24

When θ enters in an equation directly, and not as the argument of a trigonometric function, as, for example, in the case of the *spiral of Archimedes* with the equation

$$r = a\theta, \tag{29.3}$$

it is necessary to express the angle in terms of radians, since both members of the equation must have the character of length. In such cases θ takes on all possible values. Thus, if in (29.3) we take $a = 2$, and θ takes, for example, the values $0 + 2n\pi$, $\frac{\pi}{2} + 2n\pi$, $(2n+1)\pi$, $\frac{\pi}{2} + (2n+1)\pi$, n being any positive or negative integer, then the values of r are twice these values, and the reader on plotting the curve will see that it is a double spiral.

Although the same point has the two sets of coordinates (r, θ) and $(-r, \theta + 180°)$, it may be that if one of these sets of coordinates satisfies an equation and thus determines a point of the locus, the other set does not. This situation did not arise for equations (29.1) and (29.2), but it does arise for the equation $r = \cos^2 \theta$. For, if r, θ is a solution of this equation, $-r, \theta + 180°$ is not a solution.

Polar coordinates are necessarily related in some way to rectangular coordinates. A simple form of the relation is obtained when the origins of both systems coincide and the axis of the polar system is the positive x-axis of a rectangular system. In this case for r positive the relation is

$$x = r \cos \theta, \qquad y = r \sin \theta, \tag{29.4}$$

the x- and y-coordinates of a point P being the projections of the vector OP (Fig. 23) upon the respective axes. Suppose now that r is negative; then, as one sees by drawing an appropriate figure,

$$x = -|r| \cos \theta = r \cos \theta, \qquad y = -|r| \sin \theta = r \sin \theta,$$

and equations (29.4) hold in this case also.

When polar coordinates of a point are given, we obtain rectangular coordinates of the point directly from (29.4). If rectangular coordinates are given, we must solve (29.4) for r and θ. The first of these is obtained by squaring equations (29.4) and adding the results; from this we obtain

$$r = e\sqrt{x^2 + y^2}, \tag{29.5}$$

where $e = +1$ or -1. And θ must satisfy any two of the equations

$$\sin \theta = \frac{y}{e\sqrt{x^2 + y^2}}, \qquad \cos \theta = \frac{x}{e\sqrt{x^2 + y^2}}, \qquad \tan \theta = \frac{y}{x}. \tag{29.6}$$

According as we take $e = +1$ or -1 in (29.5) and (29.6), we have two sets of polar coordinates, (r, θ) and $(-r, \theta + 180°)$ for any point whose rectangular coordinates are given. If we are transforming an equation in polar coordinates into one in rectangular coordinates, we must substitute directly from (29.5) and (29.6). Then one of two things will happen: either e will enter only as e^2 and thus be replaced by $+1$, or e itself will appear. In the second case, if a polar equation admits both positive and negative values of r, there are, in fact, two equations of the locus in rectangular coordinates: one for $e = +1$ giving the points for which r is positive; the other for $e = -1$ giving the points for which r is negative. If in the second case one solves the equation for $e\sqrt{x^2 + y^2}$ and squares the resulting equation, one obtains an equation in x and y whose graph is the complete locus of the given equation in polar coordinates.

For example, on substituting from (29.5) and (29.6) in (29.1), one obtains $x^2 + y^2 = ax$, that is, a circle with center $(a/2, 0)$ and passing through the origin; this equation is an equation of the locus of (29.1).

For the equation $r = \sin\theta + \frac{1}{2}$ we have $x^2 + y^2 = y + \frac{1}{2}e\sqrt{x^2 + y^2}$. Consequently, when $e = +1$ this is an equation of the part of the curve for which r is positive; when $e = -1$, the part for which r is negative. If we solve the equation for $e\sqrt{x^2 + y^2}$ and square the result, we obtain an equation in x and y of the complete locus.

For the equation $r = \cos^2\theta$ we have $(x^2 + y^2)^{\frac{3}{2}} = x^2$, since r cannot be negative and hence $e = +1$. This equation in x and y is the equation also for $r = -\cos^2\theta$. In fact, the two polar equations are equations of the same curve, since $\cos^2(\theta + 180^\circ) = \cos^2\theta$.

When an equation of a locus is given in rectangular coordinates, its equation in polar coordinates is obtained on substituting the expressions (29.4) for x and y. Thus in polar coordinates the equation of a line $ax + by + c = 0$ is

$$r(a\cos\theta + b\sin\theta) + c = 0. \tag{29.7}$$

When $c = 0$ in (29.7), that is, when the line passes through the origin, we have for all values of r other than zero that $\tan\theta = -a/b$; and when θ satisfies this condition the equation is satisfied by all values of r; that is, $\theta = \text{const.}$ is an equation of a line through the origin. It should be remarked that in polar coordinates an equation of a line is not of the first degree in r and θ, as is the case when rectangular coordinates are used; in fact, (29.7) is not an algebraic equation in r and θ.

Since any trigonometric function is expressible algebraically in terms of x and y by means of (29.6), it follows that when an equation in polar coordinates is an algebraic function of r and of one or more trigonometric functions of θ, the corresponding equation in rectangular coordinates is algebraic.

EXERCISES

1. Draw the graph of each of the following equations for a positive, using a table of natural trigonometric functions or radian measure, as the case requires:

a. $r = a \sin \theta$.
b. $r = a(1 - \cos \theta)$ (the cardioid).
c. $r = a \sin 2\theta$.
d. $r = a \sin 3\theta$.
e. $r = a \sec 2\theta$.
f. $r^2 = 2a^2 \cos 2\theta$ (the lemniscate).
g. $r = a(\cos \theta - \sin \theta)$.
h. $r = 2a \cos \theta \pm l$ (the limaçon).
i. $r = \cos \theta - \sec^2 \theta$.
j. $r\theta = a$.
k. $r^2 = a\theta$.

2. Plot separately the portions of the curve *h* in Ex. 1 for which r is positive and r is negative; derive equations in rectangular coordinates of each portion, and an equation of the complete locus.

3. For what values of k is r always positive for the curve

$$r = \sin^2 \theta + k;$$

for what values of k is r always negative? Plot a curve of each set.

4. For which of the curves in Ex. 1 for a positive does r have positive, zero, and negative values; for which no negative values; for which no positive values?

5. Find equations in rectangular coordinates of the curves in Ex. 1.

6. What range for θ must be used for the equation $r = \sin \frac{\theta}{a}$ in order to obtain the complete graph of the equation, a being some integer?

7. Let AB be a fixed line perpendicular to the axis of a polar coordinate system meeting the axis in the point D; denote by M the point in which any vector through O meets the line. The locus of the points P and P' such that $r = OM + l$, $r' = OM - l$, where l is a fixed length, is called the *conchoid*. Find its equations in polar and also in rectangular coordinates.

8. A circle of radius a passes through O and has the polar axis for diameter OA; a line through O meets the circle in Q, and the tangent to the circle at A in R. The locus of P on the line OR such that $PR = OQ$ is called the *cissoid*. Find its equations in polar and also in rectangular coordinates.

9. Find the points of intersection of the curves $r = a \cos \theta$ and $r = a \sin \theta$; also of the curves $r = a\theta$ and $r = -a\theta$ for $\theta \geqq 0$.

30. Transformations of Rectangular Coordinates in Space

In this section we derive the equations connecting the coordinates of a point in space with reference to two different sets of rectangular axes.

We derive first the equations connecting the coordinates of a point with reference to two sets of axes respectively parallel to one another, as shown in Fig. 25. The origin O' of the $x'y'z'$-system is the point (x_0, y_0, z_0) with respect to the xyz-system, and hence the origin O of the latter system is the point $(-x_0, -y_0, -z_0)'$ of the $x'y'z'$-system. As in § 28, we obtain the following equations:

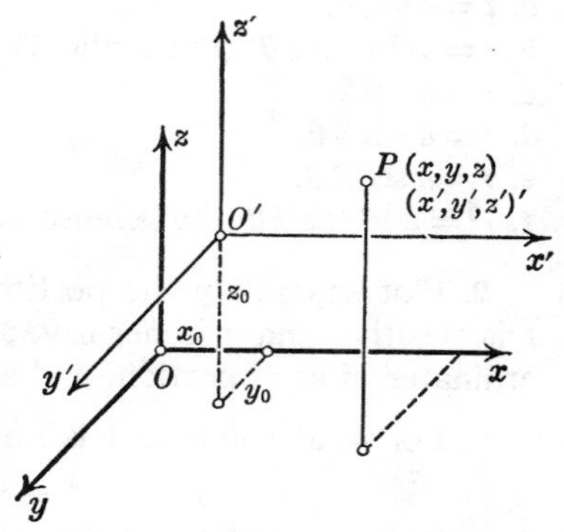

FIG. 25

$$
\begin{aligned}
x' &= x - x_0, & x &= x' + x_0, \\
y' &= y - y_0, & y &= y' + y_0, \qquad (30.1)\\
z' &= z - z_0, & z &= z' + z_0.
\end{aligned}
$$

If the expressions for x, y, and z in the second set are substituted in an equation in x, y, and z, one obtains the corresponding equation in x', y', z'. A transformation (30.1) is called a *translation* of the axes.

We consider next the general case, as shown in Fig. 26, for which the origin O' of the $x'y'z'$-system is the point $(x_0, y_0, z_0,)$ with respect to the xyz-system, and the direction cosines of the axes $O'x'$, $O'y'$, $O'z'$ with respect to the xyz-system are

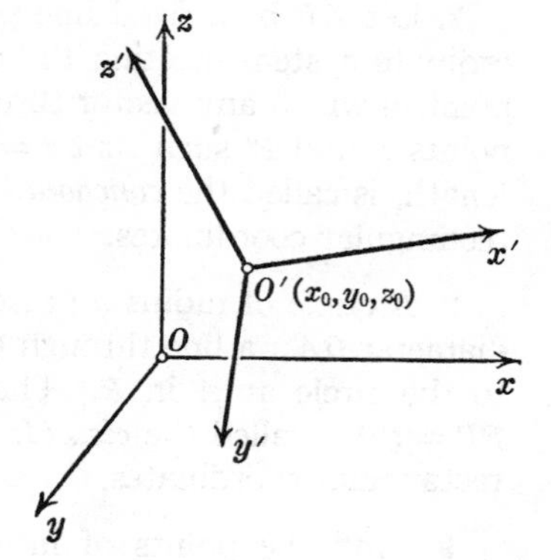

FIG. 26

$$\lambda_1, \mu_1, \nu_1; \quad \lambda_2, \mu_2, \nu_2; \quad \lambda_3, \mu_3, \nu_3$$

respectively. Being direction cosines, they must satisfy the equation (15.6); hence we have the three equations

$$(30.2) \qquad \lambda_i^2 + \mu_i^2 + \nu_i^2 = 1 \qquad (i = 1, 2, 3).$$

This notation means that one gives i the value 1 and gets one equation; then the value 2 and gets a second equation; and so on. Moreover, since the axes are mutually perpendicular, in accordance with Theorem [**16.8**] we have

$$(30.3) \qquad \begin{aligned} \lambda_1\lambda_2 + \mu_1\mu_2 + \nu_1\nu_2 &= 0, \\ \lambda_1\lambda_3 + \mu_1\mu_3 + \nu_1\nu_3 &= 0, \\ \lambda_2\lambda_3 + \mu_2\mu_3 + \nu_2\nu_3 &= 0. \end{aligned}$$

Equations (30.2) and (30.3) are the conditions that the quantities involved are direction cosines of three mutually perpendicular lines, but do not determine the mutual orientation of positive directions on these lines. Consequently we must impose a condition to ensure that the mutual orientation of the axes shall be the same in both systems, mutual orientation being as defined in § 14 and illustrated by the edges of a room meeting in a corner of the floor, the corner being the origin of the system. If we start with an xyz-system, it is evident geometrically that, once the positive axes $O'x'$, $O'y'$ of a second system are chosen, the positive axis $O'z'$ is completely determined by the prescribed mutual orientation of the axes. Suppose then that λ_1, μ_1, ν_1 and λ_2, μ_2, ν_2 are given; if we apply Theorem [**20.1**] to the second and third of equations (30.3), we obtain

$$(30.4) \qquad \begin{aligned} \lambda_3 &= t(\mu_1\nu_2 - \mu_2\nu_1), \\ \mu_3 &= t(\nu_1\lambda_2 - \nu_2\lambda_1), \\ \nu_3 &= t(\lambda_1\mu_2 - \lambda_2\mu_1). \end{aligned}$$

Substituting these expressions in (30.2) for $i = 3$, we obtain

$$(30.5) \quad t^2[(\mu_1\nu_2 - \mu_2\nu_1)^2 + (\nu_1\lambda_2 - \nu_2\lambda_1)^2 + (\lambda_1\mu_2 - \lambda_2\mu_1)^2] = 1.$$

Since the axes $O'x'$ and $O'y'$ are perpendicular by hypothesis, it follows from equation (15.10) that the expression in brackets in (30.5) is equal to $+1$, and consequently t is $+1$ or -1.

When we substitute the expressions (30.4) in the determinant

$$(30.6) \qquad D = \begin{vmatrix} \lambda_1 & \mu_1 & \nu_1 \\ \lambda_2 & \mu_2 & \nu_2 \\ \lambda_3 & \mu_3 & \nu_3 \end{vmatrix}$$

and note that the expression in brackets in (30.5) is equal to $+1$, we find that the determinant D is equal to t. When, in particular, the axes of the $x'y'z'$-system are parallel to those of the xyz-system, as in Fig. 25, the determinant (30.6) is

$$\begin{vmatrix} 1 & 0 & 0 \\ 0 & 1 & 0 \\ 0 & 0 & 1 \end{vmatrix},$$

whose value is $+1$. Consequently we impose the condition that the determinant (30.6) be $+1$, to ensure that if the axes of the $x'y'z'$-system, as shown in Fig. 26, are rotated about O', so that the axes $O'x'$ and $O'y'$ become parallel to Ox and Oy respectively, then the axis $O'z'$ is parallel to Oz and has the same sense.

When we put $t = 1$ in (30.4), we have that the right-hand members of (30.4) are the cofactors of the corresponding left-hand members in the determinant (30.6). It is readily shown that similar results hold for the other direction cosines, so that we have the theorem

[30.1] *When the direction cosines of three lines satisfy the conditions* (30.3) *and* $D = 1$ *in* (30.6), *any one of them is equal to its cofactor in the determinant* D.

With reference to the $x'y'z'$-system the direction cosines of Ox, Oy, Oz are $\lambda_1, \lambda_2, \lambda_3$; μ_1, μ_2, μ_3; ν_1, ν_2, ν_3. Consequently we have also the equations

$$(30.7) \qquad \begin{aligned} \lambda_1^2 + \lambda_2^2 + \lambda_3^2 &= 1, & \lambda_1\mu_1 + \lambda_2\mu_2 + \lambda_3\mu_3 &= 0, \\ \mu_1^2 + \mu_2^2 + \mu_3^2 &= 1, & \mu_1\nu_1 + \mu_2\nu_2 + \mu_3\nu_3 &= 0, \\ \nu_1^2 + \nu_2^2 + \nu_3^2 &= 1, & \lambda_1\nu_1 + \lambda_2\nu_2 + \lambda_3\nu_3 &= 0. \end{aligned}$$

These equations are in keeping with the above theorem, as is readily verified.

Equations of the $y'z'$-plane, the $x'z'$-plane, and the $x'y'$-plane in the xyz-coordinate system are respectively

$$
\begin{aligned}
&\lambda_1(x - x_0) + \mu_1(y - y_0) + \nu_1(z - z_0) = 0,\\
(30.8)\qquad &\lambda_2(x - x_0) + \mu_2(y - y_0) + \nu_2(z - z_0) = 0,\\
&\lambda_3(x - x_0) + \mu_3(y - y_0) + \nu_3(z - z_0) = 0.
\end{aligned}
$$

When the axes $O'x'$, $O'y'$, $O'z'$ are so placed that each of them makes an acute angle with the positive direction of the axis Oz, the direction cosines ν_1, ν_2, and ν_3 are positive; consequently we can apply directly Theorem [**18.1**] and have that the distances of the point $P(x, y, z)$ from the planes (30.8), that is, the coordinates x', y', z' in the new system, are given by

$$
\begin{aligned}
&x' = \lambda_1(x - x_0) + \mu_1(y - y_0) + \nu_1(z - z_0),\\
(30.9)\qquad &y' = \lambda_2(x - x_0) + \mu_2(y - y_0) + \nu_2(z - z_0),\\
&z' = \lambda_3(x - x_0) + \mu_3(y - y_0) + \nu_3(z - z_0).
\end{aligned}
$$

When the new axes are not so placed, the above equations hold just the same, as can be shown. For example, from the discussion in § 18 it follows that if ν in any of the equations (30.9) is negative, then below the corresponding plane is its positive side with respect to the xyz-system; thus if ν_1 is negative the positive x'-axis is directed downward. If for any position of the new axes the signs of the coefficients in two of equations (30.9) are chosen in accordance with the positive directions of the corresponding new axes relative to the xyz-system, the appropriate signs of the coefficients of the third equation are determined in accordance with Theorem [**30.1**], which assures that the positive directions of the axes in the $x'y'z'$-system have proper mutual orientation. For example, suppose $O'x'$ lies in the first, or principal, octant, in which case λ_1, μ_1, ν_1 are all positive, and $O'z'$ is tilted forward so that $O'y'$ is directed downward. Now λ_3 is negative, since $O'z'$ makes an obtuse angle with Ox, and μ_3 and ν_3 are positive. In accordance with the above theorem, $\nu_2 = \lambda_3\mu_1 - \lambda_1\mu_3$; the right-hand member of this equation is negative, and consequently ν_2 is negative, as should be the case because of the position of $O'y'$. The reader should verify the statement following (30.9) by considering the other possible cases.

If we multiply equations (30.9) by $\lambda_1, \lambda_2, \lambda_3$ respectively, add the results, and make use of (30.7), we obtain the first of the following equations, the others being obtained similarly, using μ_1, μ_2, μ_3 and ν_1, ν_2, ν_3 respectively as multipliers:

$$\begin{aligned} x &= \lambda_1 x' + \lambda_2 y' + \lambda_3 z' + x_0, \\ y &= \mu_1 x' + \mu_2 y' + \mu_3 z' + y_0, \\ z &= \nu_1 x' + \nu_2 y' + \nu_3 z' + z_0. \end{aligned} \tag{30.10}$$

These are the equations of the inverse of the transformation (30.9). When, in particular, the center O' of the $x'y'z'$-system coincides with the origin O of the xyz-system, the equations of the transformation are

$$\begin{aligned} x &= \lambda_1 x' + \lambda_2 y' + \lambda_3 z', \\ y &= \mu_1 x' + \mu_2 y' + \mu_3 z', \\ z &= \nu_1 x' + \nu_2 y' + \nu_3 z'. \end{aligned} \tag{30.11}$$

The transformation (30.11) is sometimes referred to as that corresponding to a *rotation* of the original axes. When the expressions (30.10), or (30.11), for x, y, and z are substituted in an equation (or equations) of a locus referred to the xyz-system, the resulting equation (or equations) is an equation (or are equations) of the locus in the $x'y'z'$-system.

The inverse of a transformation (30.11) is

$$\begin{aligned} x' &= \lambda_1 x + \mu_1 y + \nu_1 z, \\ y' &= \lambda_2 x + \mu_2 y + \nu_2 z, \\ z' &= \lambda_3 x + \mu_3 y + \nu_3 z, \end{aligned} \tag{30.12}$$

as follows from (30.9) on putting x_0, y_0, z_0 equal to zero.

The reader should observe that in any of the equations (30.11) and (30.12) the coefficient of any term on the right is the cosine of the angle between the axis of the coordinate which is multiplied by the coefficient considered and the axis of the coordinate on the left-hand side of the equation. For example, from the second of (30.11), and also from the third of (30.12) we have that μ_3 is the cosine of the angle between Oy and Oz'. This observation and equations (30.11) and (30.12) are set forth in the following table:

	x'	y'	z'
x	λ_1	λ_2	λ_3
y	μ_1	μ_2	μ_3
z	ν_1	ν_2	ν_3

Any element in the square is the cosine of the angle between the axes in whose row and column it lies. Moreover, an equation (30.12) is obtained by multiplying each element in the same column as x', y', or z' by the coordinate in the same row on the left, adding the results and equating this to the coordinate at the top of the column. Equations (30.11) are similarly obtained by using rows instead of columns.

EXERCISES

1. Transform by a suitable transformation (30.1) the equation $x^2 + 4y^2 + 3z^2 - 2x - 16y + 12z + 28 = 0$ so that in the resulting equation there are no terms of the first degree. Is this possible for the equation $xy + 2z - 3 = 0$?

2. Show that for the second of transformations (30.1), and also for (30.11), the expression (15.2) is transformed into an expression of the same form in the new coordinates. Why should one expect this to be the case?

3. Show that the planes $x + 2y + 2z = 0$, $2x + y - 2z = 0$, $2x - 2y + z = 0$ may be used as coordinate planes of a coordinate system, and find the equations of the corresponding transformation.

4. Show that the three planes $x + y + z - 4 = 0$, $x - 2y + z + 2 = 0$, and $x - z = 0$ may be taken as the coordinate planes of a rectangular coordinate system, and find the corresponding transformation of coordinates.

5. Find the equations of a transformation of coordinates so that the plane $x + y + z = 0$ is the $x'y'$-plane in the new system.

6. Transform by means of (30.11) the equation of the sphere $x^2 + y^2 + z^2 + 2fx + 2gy + 2hz + e = 0$ and discuss the result.

7. Interpret the transformation

$$x = x' \cos\theta - y' \sin\theta, \quad y = x' \sin\theta + y' \cos\theta, \quad z = z'$$

as a particular type of rotation of the axes.

8. How must θ be chosen in a transformation of the type of Ex. 7 so that the equation $ax^2 + by^2 + cz^2 + 2\,hxy + d = 0$ may be transformed into one lacking a term in $x'y'$?

9. Find equations of the line each point of which has the same coordinates in two coordinate systems in the relation (30.11); interpret the result geometrically.

10. Show that three lines with direction numbers 1, 2, 2; -2, -1, 2; and 2, -2, 1 are mutually perpendicular, and find the transformation of coordinates to an $x'y'z'$-system having lines with these direction numbers and through the point (2, 0, -3) for coordinate axes. Apply this transformation to the equation

$$4\,x^2 + 4\,y^2 - 8\,z^2 + xy - 5\,xz - 5\,yz + 9 = 0.$$

11. Show that the equations

$$\begin{aligned} x = x'\,(&\cos\phi\cos\psi - \sin\phi\sin\psi\cos\theta) \\ &- y'(\cos\phi\sin\psi + \sin\phi\cos\psi\cos\theta) + z'\sin\phi\sin\theta, \\ y = x'\,(&\sin\phi\cos\psi + \cos\phi\sin\psi\cos\theta) \\ &- y'(\sin\phi\sin\psi - \cos\phi\cos\psi\cos\theta) - z'\cos\phi\sin\theta, \\ z = x'&\sin\psi\sin\theta + y'\cos\psi\sin\theta + z'\cos\theta, \end{aligned}$$

for any values of θ, ϕ, and ψ, are equations of a transformation (30.11). These equations are known as *Euler's formulas.*

31. Spherical and Cylindrical Coordinates

In the study of certain phenomena in space there are systems of coordinates other than rectangular coordinates which are found to be more useful. We define one such system and establish its relation to a rectangular system, as shown in Fig. 27. The position of a point P is determined by its distance r from a point O, by the angle ϕ which the vector OP makes with a fixed vector Oz, and by the angle θ which the plane of the

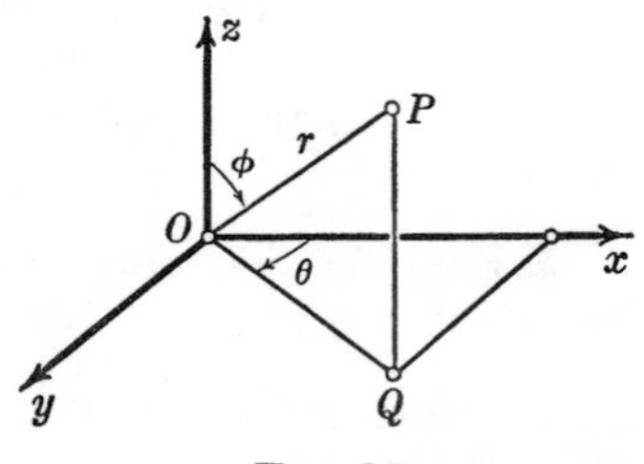

FIG. 27

vectors OP and Oz makes with a fixed plane through Oz. When O is taken for origin, Oz as the positive z-axis, and the fixed plane through Oz as the xz-plane of a rectangular coordinate system, the angle θ is the angle between the x-axis and the projection of OP on the xy-plane. We make the convention that θ is measured from Ox toward Oy. Hence we have the equations

$$x = r \sin \phi \cos \theta, \quad y = r \sin \phi \sin \theta, \quad z = r \cos \phi. \tag{31.1}$$

Squaring these equations and adding the results, we find that

$$r = \sqrt{x^2 + y^2 + z^2}. \tag{31.2}$$

Then from (31.1) we have as the other equations of the inverse transformation

$$\cos \phi = \frac{z}{\sqrt{x^2 + y^2 + z^2}}, \qquad \tan \theta = \frac{y}{x}. \tag{31.3}$$

The coordinates r, ϕ, and θ as defined are called *spherical coordinates*; some writers call them *polar coordinates in space.*

From (31.2) it follows that the surfaces $r = k$ as the constant k takes different values are spheres with O as their common center. If we think of O as the center of the earth, of the line joining the center to the North Pole as the fixed vector of reference Oz, and the plane through this line and Greenwich as the fixed plane of reference, then θ is the west longitude of a point on the earth's surface, and $90° - \phi$ is its latitude; ϕ is sometimes called the *colatitude* of the point.

Another system of spatial coordinates is defined by the equations

$$x = r \cos \theta, \quad y = r \sin \theta, \quad z = d, \tag{31.4}$$

from which we have as the inverse

$$r = \sqrt{x^2 + y^2}, \quad \tan \theta = \frac{y}{x}, \quad d = z. \tag{31.5}$$

In this system the surfaces $r = k$ as the constant k takes different values are cylinders with the z-axis as common axis. The quantities r, θ, d defined by (31.5) are called *cylindrical coordinates.* These equations have for basis a plane with a polar coordinate system, and an axis perpendicular to the plane at the pole of the polar system.

EXERCISES

1. What are the surfaces $\theta = \text{const.}$ in spherical coordinates? What are the surfaces $\phi = \text{const.}$?

2. Discuss the several loci defined by the equations obtained when two of the spherical coordinates are equated to constants.

3. Find the direction cosines with respect to the xyz-system of the line from the origin to the point whose spherical coordinates are r, ϕ, θ.

4. Find the expression for the distance between two points in spherical coordinates; in cylindrical coordinates.

5. Find the spherical coordinates of the points whose rectangular coordinates are $(2, 4, 3)$, $(3, 3, -2)$, $(-1, 2, -2)$.

6. Find the cylindrical coordinates of the points in Ex. 5.

7. Find equations in spherical coordinates of the loci whose equations in rectangular coordinates are

a. $x^2 + y^2 = 5$.

b. $3x - 4y + 5z - 1 = 0$.

c. $4x^2 - y^2 = 1$.

d. $x^2 + 2y^2 + 3z^2 - 6 = 0$.

e. $xy + yz + xz = 0$.

CHAPTER 4

The Conics. Locus Problems

II

32. A Geometric Definition of the Conics

In the preceding chapters we have obtained in a number of cases an equation of the locus of a point which satisfies a certain relation to points and lines defined with reference to a given coordinate system, and we have given certain exercises to find a locus so defined. An equation of such a locus depends for its form not only upon the geometric character of the locus but also upon the particular coordinate system used. In view of the results of § 28 it is clear that in finding an equation of a given locus one is free to choose a coordinate system with respect to which the fixed objects (points, lines, etc.) involved in the definition of the locus have such position that the equation of the locus, and the calculation involved, shall be as simple as possible, provided that in so doing one is not imposing additional conditions upon the locus. For example, the results of Ex. 21 of § 13 hold for any triangle, because a rectangular coordinate system can be chosen so that the coordinates of the vertices of any triangle are as given in this exercise. On the other hand, the results of Ex. 20 of § 13 apply only to a right-angled triangle if we are using rectangular axes, but the results would be true for any triangle if shown to be true for a triangle with vertices $O(0, 0)$, $A(a, 0)$, $B(b, c)$. We remark that a coordinate system can be chosen so that any two points have the coordinates $(b, 0)$ and $(-b, 0)$ by taking the line through the points for the x-axis and the origin at the mid-point of the segment connecting the two points; accordingly the geometric character of the locus in Ex. 9 of § 13 is not changed, but its equation is simplified, by taking $(b, 0)$ and $(-b, 0)$ as the two points.

We make use of this idea of choice of an advantageous coordinate system in deriving equations of the conics from the following geometric definition:

Let F be a fixed point, called the focus, and d a fixed line, not through the point, called the directrix; then a conic is the locus of a point P such that the ratio of the distance FP to the distance of P from the line d is a positive constant, e, called the eccentricity; it is a parabola when $e = 1$, an ellipse when $e < 1$, and a hyperbola when $e > 1$.

Distance as used in this definition is numerical, or absolute, distance. In § 38 we show the equivalence of this definition and the definition of the conics as plane sections of a right circular cone.

From the above definition it follows that the quantities which determine, and therefore can affect, the size and shape of a conic are its eccentricity and the distance of the focus from the directrix. Hence conics of all possible shapes and sizes are obtained by taking all values of the eccentricity, a given line d for directrix, and a given line f perpendicular to d for a line of foci, a particular focus being determined by its distance from d. If two other lines d' and f' are chosen for directrix and line of foci respectively, a conic defined with respect to these lines can be brought into coincidence with one of the set of conics defined with respect to d and f by bringing d' into coincidence with d and the focus into coincidence with one of the points of f.

In order to obtain an equation of the conics, we take the directrix as the y-axis and the perpendicular to the directrix through the focus as the x-axis, and denote the focus by $(k, 0)$. The distances of a representative point (x, y) from the focus and directrix are $\sqrt{(x-k)^2+y^2}$ and $|x|$ respectively. In accordance with the definition, we have

$$\text{(32.1)} \qquad \frac{FP}{DP} = \frac{\sqrt{(x-k)^2+y^2}}{|x|} = e,$$

where D denotes the foot of the perpendicular from P upon the directrix. From the equation

$$\sqrt{(x-k)^2+y^2} = e\,|x|,$$

upon squaring both sides, we obtain

$$\text{(32.2)} \qquad (1-e^2)x^2 - 2\,kx + y^2 + k^2 = 0.$$

Suppose now that we consider in connection with equation (32.2) an equation of a conic of eccentricity e with the y-axis for directrix and for focus the point $(tk, 0)$, where t is a positive constant. If we denote by $P'(x', y')$ a representative point on the conic, the equation of the conic is found to be

$$(1 - e^2)x'^2 - 2\,tkx' + y'^2 + t^2k^2 = 0. \tag{32.3}$$

If then we substitute

$$x' = tx, \qquad y' = ty \tag{32.4}$$

in this equation, we obtain equation (32.2) multiplied by t^2. Hence to each point on either curve there corresponds a point on the other such that the line joining corresponding points passes through the origin, that is, through the point of intersection of the directrix and the perpendicular upon the latter through the focus. Moreover, the distances of P and P' from the origin are related as follows:

$$OP' = \sqrt{x'^2 + y'^2} = t\sqrt{x^2 + y^2} = t \cdot OP.$$

Thus if $t > 1$ the second curve is a magnification of the first, and vice versa if $t < 1$. In both cases, as a matter of convenience, we say that either curve is a *magnification* of the other.

Two curves in a plane are said to be *similar* when there is a point O in the plane such that if P is any point on one curve, the line OP meets the other curve in a point P', and the ratio OP/OP' is a constant for all such lines. Also two curves are said to be *similar* when either curve is congruent to a suitable magnification of the other. Likewise in space two figures of any kind are said to be similar if their points are related to a point O as above, or if either is congruent to a suitable magnification of the other.

As a result of the discussion of equations (32.2) and (32.3) and the definition of similar curves, we have

[32.1] *Two conics having the same eccentricity are similar; in particular, all parabolas are similar.*

By a suitable translation of the axes we shall obtain in § 33 an equation of the parabola, and in § 35 equations of the ellipse and the hyperbola, in simpler form than (32.2). When the equations are in these simpler forms, the geometric properties of the loci are more readily obtained.

EXERCISES

1. Show that in accordance with the definition of a conic the line through the focus perpendicular to the directrix is an *axis of symmetry* of the curve; that is, if P is any point on the conic, so also is the point symmetric to P with respect to this axis (see § 2). Show also geometrically that there is one and only one point of a parabola on this axis, and that there are two and only two points of an ellipse or hyperbola on this axis. Verify these statements algebraically from equation (32.2).

2. Find an equation of the conics when the directrix is taken as the x-axis, and the line perpendicular to the directrix through the focus for the y-axis, and the focus is denoted by $(0, k)$.

3. Find an equation of the parabola whose directrix is the line $x - 1 = 0$ and whose focus is $(2, 3)$; also an equation of an ellipse of eccentricity $\frac{1}{2}$ with this directrix and focus.

4. Find an equation of a hyperbola of eccentricity 2 whose directrix is the line $4x - 3y + 2 = 0$ and whose focus is $(1, -1)$.

5. Find an equation of the conic of eccentricity e whose directrix is the line $ax + by + c = 0$ and whose focus is the point (h, k), and reduce the equation to the form

$$Ax^2 + 2Hxy + By^2 + 2Fx + 2Gy + C = 0.$$

6. Given two triangles whose sides are proportional, show that a point O can be found with respect to either triangle such that a suitable magnification of this triangle with respect to O gives a triangle congruent to the other.

33. The Parabola

Since a parabola is a conic for which the eccentricity e is $+1$, equation (32.2) for a parabola may be written

$$y^2 = 2k\left(x - \frac{k}{2}\right). \tag{33.1}$$

In order to obtain an equation in simpler form, we make the translation of the axes defined by

$$x' = x - \frac{k}{2}, \qquad y' = y, \tag{33.2}$$

and equation (33.1) is transformed into

$$y'^2 = 2kx'.$$

In this coordinate system the *axis* of the parabola, that is, the line through the focus perpendicular to the directrix, is the x'-axis, and the point of the parabola on its axis, called the *vertex*, is at the origin. Moreover, the focus, which is $(k, 0)$ in the xy-system, is $(k/2, 0)'$ in the $x'y'$-system, as follows from (33.2); and similarly, an equation of the directrix, which is $x = 0$ in the xy-system, is $x' + \frac{k}{2} = 0$ in the $x'y'$-system.

As we shall use only this new coordinate system in what follows, we use x and y to denote the coordinates instead of x' and y'. If then, in order to avoid fractions, we put $k = 2a$, an equation of the parabola is

$$y^2 = 4ax, \tag{33.3}$$

its focus is the point $(a, 0)$, and its directrix the line $x + a = 0$. This is shown in Fig. 28 for the case when a is positive.

The perpendicular to the axis at the focus, that is, the line $x - a = 0$, meets the parabola in the points $(a, 2a)$ and $(a, -2a)$. The line segment with these points as end points is called the *latus rectum* of the parabola; its length is the numerical value of $4a$.

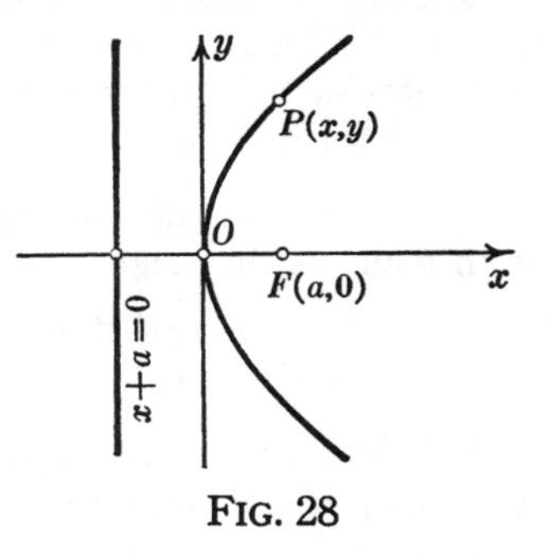

FIG. 28

Since the y-axis meets the parabola in the origin counted doubly, it is the *tangent* to the parabola at its vertex. From (33.3) it is seen that a parabola is characterized geometrically as follows:

[33.1] *A parabola is the locus of a point P the square of whose distance from a line (its axis) is equal to four times the product of the directed distances of P and the focus of the parabola from its tangent at the vertex, this tangent being the perpendicular to the axis at the vertex.*

In consequence of this theorem an equation of a parabola is determined by equations of its axis and of the tangent at the vertex, and the distance of its focus from this tangent (see Exs. 7, 8).

EXERCISES

1. Find the focus and directrix of the parabola $y^2 = 8x$, and the points in which it is intersected by each of the lines

$$2x - y - 1 = 0, \qquad 2x - y + 1 = 0.$$

2. Draw the graph of each of the following parabolas, and its focus and directrix:

a. $y^2 = 12x$. *c.* $x^2 = 8y$.
b. $y^2 = -4x$. *d.* $x^2 = -12y$.

3. Denoting by V and P the vertex and a representative point on a parabola, show that if P' is the point of the line containing VP such that $VP' = tVP$, where t is any constant not equal to zero, the locus of P' is a parabola; and determine the relation between the foci and directrices of the two parabolas.

4. Find an equation of the parabola whose directrix is the x-axis and whose focus is the point (0, 4), and find a transformation of coordinates so that in the new $x'y'$-system the equation of the parabola is of the form (33.3).

5. Prove that each of the following is an equation of a parabola with focus at the origin:

a. $y^2 = 4a(x + a)$. *c.* $y^2 = -4a(x - a)$.
b. $x^2 = 4a(y + a)$. *d.* $x^2 = -4a(y - a)$.

6. Find an equation of the parabola with vertex at the point (1, 2) and focus at the point (– 1, 2).

7. Given the lines $y - 4 = 0$ and $x + 3 = 0$, find an equation of the parabola of latus rectum 8 which has these lines for axis and tangent at the vertex, when the first line is the axis and the curve is to the right of the second line; when the second line is the axis and the curve is below the first line.

8. Find an equation of the parabola with the lines $x - 2y + 1 = 0$ and $2x + y - 3 = 0$ for axis and tangent at the vertex respectively, with latus rectum of length 6, and which lies below the tangent at the vertex.

9. Find an equation of the parabola whose directrix is the line $3x - 4y - 1 = 0$ and whose focus is (1, – 2). What are equations of its axis and of the tangent at its vertex?

34. Tangents and Polars

The coordinates of the points of intersection of the line

$$y = mx + h \tag{34.1}$$

with the parabola (33.3) are found by solving their equations simultaneously. Substituting the expression (34.1) for y in (33.3) and collecting terms, we get

$$m^2x^2 + 2(mh - 2a)x + h^2 = 0 \tag{34.2}$$

as the equation whose roots are the x-coordinates of the points of intersection. When $m = 0$, that is, when the line is parallel to the axis of the parabola, there is one point of intersection, namely $\left(\frac{h^2}{4a}, h\right)$. When $m \neq 0$, we have from (34.2)

$$x = \frac{2a - mh \pm 2\sqrt{a(a - mh)}}{m^2}. \tag{34.3}$$

According as the quantity under the radical is positive or negative, there are two real points of intersection or none; in the latter case we say that the points of intersection are *conjugate imaginary*.

When the line (34.1) intersects the parabola in two real points, the x-coordinate of the mid-point of the segment joining these points, being one half the sum of the two values of x in (34.3), is $(2a - mh)/m^2$; and the y-coordinate of the mid-point is found, on substituting this value of x in (34.1), to be $2a/m$. Since this value does not depend upon the value of h in the equation (34.1) of the line, we have

[34.1] *The mid-points of a set of parallel chords of a parabola lie on a line parallel to the axis of the parabola.*

When the quantity under the radical in (34.3) is equal to zero, that is, when $h = a/m$, the two points of intersection coincide and the line is tangent to the parabola at the point $\left(\frac{a}{m^2}, \frac{2a}{m}\right)$ (see § 12 following equation (12.8)). Hence we have

[34.2] *For every value of m different from zero the line*

$$y = mx + \frac{a}{m} \tag{34.4}$$

is tangent to the parabola $y^2 = 4\,ax$, the point of tangency being $\left(\frac{a}{m^2}, \frac{2\,a}{m}\right)$.

The tangent (34.4) meets the x-axis (see Fig. 29) in the point $T(-\,a/m^2, 0)$, and the length of the segment TF of the axis is $a\left(1 + \frac{1}{m^2}\right)$. The distance from the focus F to the point of contact $P(a/m^2, 2\,a/m)$ is of the same length, as is readily shown. Hence FTP is an isosceles triangle with vertex at F. Since the tangent makes equal angles with the x-axis and any line parallel to it, we have

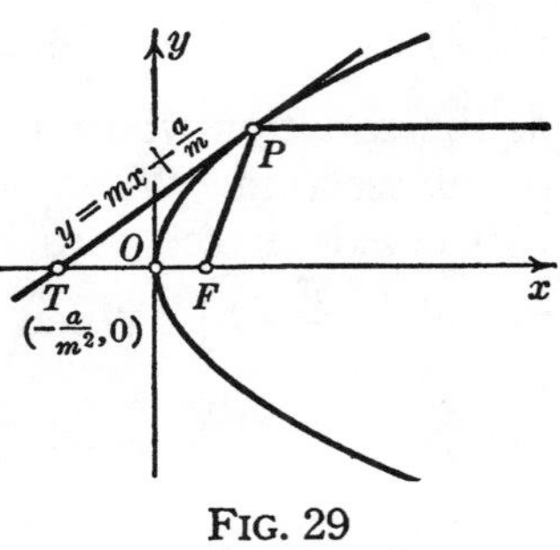

FIG. 29

[34.3] *The tangent to a parabola at any point P makes equal angles with the line joining the focus to P and with the line through P parallel to the axis of the parabola.*

When a parabola is revolved about its axis, the surface so generated is called a *paraboloid of revolution*, and the axis of the generating parabola is called the *axis* of the paraboloid. Each plane section of the surface by a plane containing the axis is a parabola, and the parabolas have the same focus. A mirror with such a curved surface is called parabolic. Recalling from physics that the angle of reflection from a mirror is equal to the angle of incidence, we have from Theorem [**34.3**] that rays of light parallel to the axis of a parabolic mirror are reflected to the focus, and, conversely, when a light is placed at the focus of a parabolic mirror, the rays emanating from it, upon reflection by the mirror, emerge parallel to the axis. This phenomenon is given practical use in reflecting telescopes and automobile headlights.

If we multiply equation (34.4) by $2\,a/m$ and write the result in the form

$$\frac{2\,a}{m}y = 2\,a\left(x + \frac{a}{m^2}\right),$$

this equation of the tangent becomes in terms of the point of contact $x_1 = a/m^2$, $y_1 = 2a/m$

$$y_1 y = 2a(x + x_1). \tag{34.5}$$

When (x_1, y_1) is any point of the plane, equation (34.5) is an equation of a line. The line so defined is called the *polar* of the point (x_1, y_1) with respect to the parabola, and the point (x_1, y_1) is called the *pole* of the line. In particular, the polar of a point on the parabola is the tangent at the point. We shall now find the geometric significance of the polar of a point not on the parabola.

If we solve (34.5) for x and substitute the result in (33.3), we obtain the quadratic

$$y^2 - 2y_1 y + 4ax_1 = 0,$$

whose two roots are the y-coordinates of the points of intersection of the line and the parabola. These roots are

$$y = y_1 \pm \sqrt{y_1^2 - 4ax_1}.$$

Hence the line meets the parabola in two real points or in two conjugate imaginary points according as $y_1^2 - 4ax_1$ is a positive or negative number, that is, according as the point (x_1, y_1) lies outside or inside the parabola. To prove the latter statement we observe that when a is positive and x_1 is negative, the quantity under the radical is positive and the point $P_1(x_1, y_1)$ lies to the left of the y-axis, which, as previously shown, is the tangent to the parabola at its vertex, and consequently the point lies outside the parabola. For x_1 positive the line $x - x_1 = 0$ meets the parabola in two points (x_1, y_2) and $(x_1, -y_2)$, where $y_2^2 - 4ax_1 = 0$, and consequently $y_1^2 - 4ax_1$ is positive or negative according as y_1^2 is greater or less than y_2^2, that is, according as P_1 is outside or inside the parabola.

If (x_2, y_2) is any point on the polar of (x_1, y_1), we have

$$y_1 y_2 = 2a(x_2 + x_1).$$

In consequence of the fact that x_1, y_1 and x_2, y_2 enter symmetrically in this equation, it follows that the point (x_1, y_1) is on the polar of the point (x_2, y_2), namely, the line

$$y_2 y = 2a(x + x_2).$$

Hence we have

[34.4] *If the polar of a point* (x_1, y_1) *with respect to a parabola passes through the point* (x_2, y_2), *the polar of* (x_2, y_2) *passes through the point* (x_1, y_1).

By means of this theorem we derive the geometric significance of the polar of a point (x_1, y_1) lying outside the parabola. It has been shown that the polar of this point intersects the parabola in two real points. By the above theorem (x_1, y_1) lies on the polars of these two points, but these polars are the tangents to the parabola at these points. Hence we have (see Ex. 13)

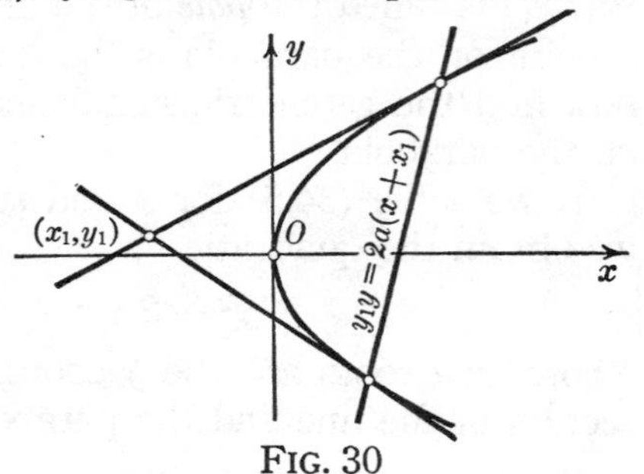

FIG. 30

[34.5] *The polar of a point outside a parabola is the line joining the points of tangency of the two tangents to the parabola through the given point.*

We have just shown indirectly that through any point (x_1, y_1) outside a parabola two tangents to the parabola can be drawn. In order to obtain equations of these tangents, we substitute x_1, y_1 for x, y in (34.4) and obtain the quadratic

$$m^2x_1 - my_1 + a = 0,$$

whose two roots are the slopes of the two tangents. These roots are

$$m = \frac{y_1 \pm \sqrt{y_1{}^2 - 4\,ax_1}}{2\,x_1},$$

which are real and distinct when the point (x_1, y_1) is outside the parabola. Substituting these values of m in (34.4), we obtain the desired equations of the two tangents.

For example, for the parabola $y^2 = 12\,x$ and the outside point $(-1, 2)$ the values of m are 1 and -3, and equations of the respective tangents are $x - y + 3 = 0$ and $3\,x + y + 1 = 0$.

EXERCISES

1. Find an equation of the tangent with the slope 3 to each of the parabolas

$$y^2 = 8x, \quad y^2 = -4x.$$

Also find equations of the tangents to the first parabola at the points for which $x = 2$, and of the tangents to the second parabola from the point (3, 2), and the angles between the tangents in the latter two cases.

2. Find an equation of the tangent to the parabola $y^2 = 5x$ which is perpendicular to the line $3x + 2y - 1 = 0$; also the coordinates of the point of contact.

3. The line segment joining the focus of a parabola to any point P of the parabola is called the *focal radius* of P. Find the coordinates of the points of the parabola (33.3) whose focal radii are equal in length to the latus rectum.

4. Show that

$$x^2 = 4ay$$

is an equation of a parabola with the y-axis for the axis of the parabola and the x-axis for tangent at the vertex. What are the coordinates of the focus and an equation of the directrix? Show that the tangent with the slope m is

$$y = mx - am^2,$$

and the point of contact is $(2am, am^2)$.

5. For what value of c is the line $2x + 3y + c = 0$ tangent to the parabola $x^2 = -6y$?

6. Show that a tangent to a parabola at a point P meets its axis produced at a point whose distance from the vertex is equal to the distance of P from the tangent to the parabola at the vertex.

7. The *normal* to a curve at a point P is by definition the perpendicular to the tangent at P. Show that an equation of the normal to the parabola (33.3) at the point (x_1, y_1) is

$$2a(y - y_1) + y_1(x - x_1) = 0.$$

8. Show that the normal to a parabola at a point P meets the axis of the parabola in a point whose distance from the projection of P upon the axis is one half the latus rectum of the parabola. This segment of the axis is called the *subnormal* for the point.

9. Show that the point of intersection of a tangent to a parabola and the perpendicular to the tangent through the focus lies on the tangent to the parabola at the vertex.

10. Show that the segment of a tangent to a parabola between the point of tangency and the directrix subtends a right angle at the focus.

11. Show that the tangents to a parabola at the extremities of any chord through the focus meet at a point on the directrix. What relation does this result bear to Theorems [**34.4**] and [**34.5**]?

12. Show that the chord joining the points of contact of any two mutually perpendicular tangents to a parabola passes through the focus.

13. Let l_1 and l_2 be two chords of a parabola through a point P within the parabola, and let P_1 and P_2 be the points of intersection of the tangents to the parabola at the extremities of l_1 and l_2 respectively. Show that the line P_1P_2 is the polar of P.

35. Ellipses and Hyperbolas

In accordance with the definition in § 32 the locus of a point P whose distance from a fixed point, the focus, is equal to a positive constant e times its distance from a fixed line, the directrix, is an ellipse when $e < 1$ and a hyperbola when $e > 1$. In § 32, on taking the y-axis for the directrix and the point $(k, 0)$ on the x-axis for focus, we obtained the equation

$$(1 - e^2)x^2 - 2\,kx + y^2 = -\,k^2. \tag{35.1}$$

If we divide this equation by $(1 - e^2)$ and complete the square of the terms in x by adding $k^2/(1 - e^2)^2$ to both sides of the equation, the result may be written

$$\left(x - \frac{k}{1 - e^2}\right)^2 + \frac{y^2}{1 - e^2} = \frac{k^2e^2}{(1 - e^2)^2}.$$

If, in order to obtain simpler forms of equations of ellipses and hyperbolas, we make the translation of the axes defined by

$$x' = x - \frac{k}{1 - e^2}, \quad y' = y, \tag{35.2}$$

an equation of the conic in terms of x' and y' is

$$x'^2 + \frac{y'^2}{1 - e^2} = \frac{k^2e^2}{(1 - e^2)^2}. \tag{35.3}$$

The focus and directrix, which in the xy-system are the point $(k, 0)$ and the line $x = 0$, are given in the $x'y'$-system by

$$\left(\frac{ke^2}{e^2 - 1}, 0\right)', \quad x' + \frac{k}{1 - e^2} = 0 \tag{35.4}$$

respectively, as follows from (35.2).

From the form of equation (35.3) it is seen that the x'- and y'-axes are axes of symmetry of the conic (see § 2); that is, if (x', y') is a point of the conic, so also are $(x', -y')$ and $(-x', y')$. This means geometrically that the two points in which a line parallel to the y'-axis meets the conic are at equal distances from the x'-axis, and on opposite sides of the latter; and likewise the two points in which a line parallel to the x'-axis meets the conic are at equal distances from the y'-axis, and on opposite sides of the latter. Also it follows from (35.3) that the x'-axis meets the conic in the two points $\left(-\frac{ke}{1 - e^2}, 0\right)'$ and $\left(\frac{ke}{1 - e^2}, 0\right)'$; the points with these coordinates are the end points of a line segment of which the origin of the $x'y'$-system is the mid-point. Hence an ellipse or a hyperbola has two axes of symmetry, called the *principal axes*, the one (the x'-axis) being perpendicular to the directrix, and the other (the y'-axis) being the line parallel to the directrix and through the mid-point of the segment of the former axis (the x'-axis) whose end points are points of the conic.

It is seen also from (35.3) that if (x', y') is a point of the conic, so also is $(-x', -y')$; that is, any chord of the conic through the origin of the $x'y'$-system is bisected by the origin. Consequently the intersection of the principal axes of an ellipse or of a hyperbola is a point of symmetry of the curve. It is called the *center* of the conic. Since there is no center of a parabola in this sense, ellipses and hyperbolas are called *central conics.*

If, in order to simplify the equation (35.3), we define the quantity a by

$$a = \frac{ke}{1 - e^2}, \tag{35.5}$$

and drop the primes in equation (35.3), that is, use x and y for coordinates in the new system (since only this new system is used in what follows), this equation may be written

$$\frac{x^2}{a^2} + \frac{y^2}{a^2(1 - e^2)} = 1. \tag{35.6}$$

From this equation it follows that the curve meets the axis through the focus (the x-axis) in the two points $A'(-a, 0)$ and $A(a, 0)$. Also from (35.4) it follows that the coordinates of the focus and an equation of the directrix are respectively

$$F'(-ae, 0), \qquad x + \frac{a}{e} = 0. \tag{35.7}$$

Since, as remarked before, an ellipse or a hyperbola is symmetric with respect to its principal axes, which in the present coordinate system are the x- and y-axes, it follows from considerations of symmetry that for each of these curves there is a second focus and a second directrix symmetric with respect to the y-axis to F' and the directrix $x + \frac{a}{e} = 0$ respectively. They are given by

$$F(ae, 0), \qquad x - \frac{a}{e} = 0 \tag{35.8}$$

respectively. As a further proof of this result, we observe that an equation of the conic with this focus and directrix is

$$(x - ae)^2 + y^2 = e^2\left(x - \frac{a}{e}\right)^2,$$

which reduces to (35.6), as the reader should verify.

Thus far we have treated ellipses and hyperbolas simultaneously, but in proceeding further it is advisable at times to treat them separately. In such cases the treatment will be given in parallel columns, that on the left pertaining to the ellipse and that on the right to the hyperbola, while statements which apply equally to both types of conics will not be separated.

The quantity $a^2(1 - e^2)$ appearing in equation (35.6)

is positive for an ellipse, since $e < 1$. Consequently a real number b is defined by

$$b^2 = a^2(1 - e^2), \tag{35.9}$$

is negative for a hyperbola, since $e > 1$. Consequently a real number b is defined by

$$b^2 = -a^2(1 - e^2), \tag{35.9'}$$

in terms of which, as follows from (35.6), equations of the respective curves are

$$\frac{x^2}{a^2} + \frac{y^2}{b^2} = 1, \tag{35.10}$$

$$\frac{x^2}{a^2} - \frac{y^2}{b^2} = 1, \tag{35.10'}$$

in which it is understood that a and b are positive numbers.

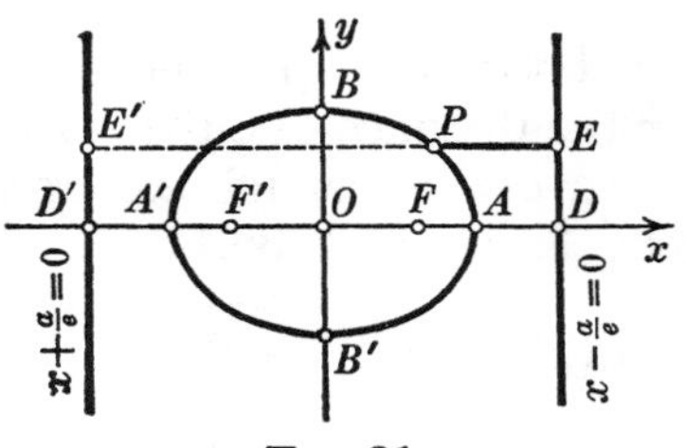

FIG. 31

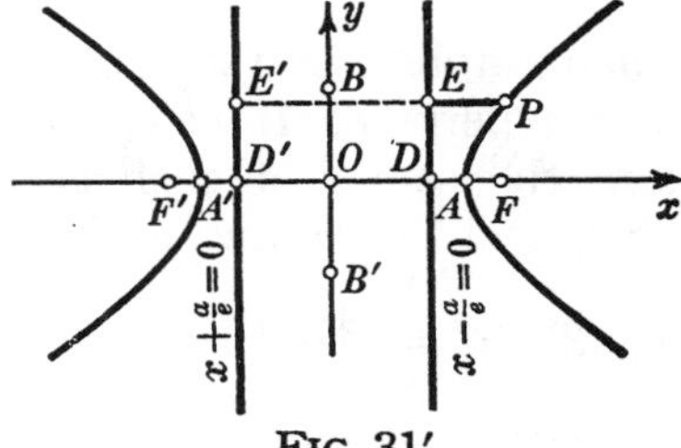

FIG. 31′

From (35.10) it follows that an ellipse meets the y-axis in the points $B'(0, -b)$ and $B(0, b)$. Also it follows from (35.9) that $b < a$. Accordingly the line segment $A'A$ of the principal axis containing the foci is called the *major axis*, and the segment $B'B$ of the other principal axis, that is, the perpendicular to the major axis through the center of the ellipse, the *minor axis*. The numbers a and b are called the *semi-major* and *semi-minor axes* respectively, in the sense that they are the lengths of these segments. The points A' and A, the extremities of the major axis, are called the *vertices* of the ellipse. (See Fig. 31.)

From (35.10′) it follows that a hyperbola does not meet the y-axis. The segment $A'A$ of the principal axis containing the foci is called the *transverse axis* of the hyperbola. The segment with end points $B'(0, -b)$ and $B(0, b)$ of the other principal axis, that is, the line through the center perpendicular to the transverse axis, is called the *conjugate axis* of the hyperbola. The numbers a and b are called the *semi-transverse* and *semi-conjugate axes* respectively. The points A' and A, the extremities of the transverse axis, are called the *vertices* of the hyperbola. The geometric significance of the end points B' and B of the conjugate axis is given in § 37. (See Fig. 31′.)

From equations (35.9) and (35.9′) we have that in the respective cases the eccentricity is expressed in terms of the semi-axes as follows:

(35.11) $$e^2 = \frac{a^2 - b^2}{a^2},$$ | (35.11′) $$e^2 = \frac{a^2 + b^2}{a^2},$$

which emphasizes the fact that $e < 1$ for an ellipse and $e > 1$ for a hyperbola. If we denote by c the distance of either focus from the center of each curve, then $c = ae$, as follows from (35.7) and (35.8), and we have from (35.11) and (35.11′)

(35.12) $c^2 = a^2 - b^2$. | (35.12′) $c^2 = a^2 + b^2$.

When these results are considered in connection with Figs. 31 and 31′, we see that

a circle with center at either end of the minor axis of an ellipse and radius equal to the semi-major axis meets the major axis in the two foci.

a circle with center at the center of a hyperbola and radius equal to the hypotenuse of a right triangle whose legs are the semi-transverse and semi-conjugate axes meets the transverse axis in the two foci (see Fig. 35).

We shall now find an important relation between the lengths of the focal radii of any point on these curves, that is, the lengths $F'P$ and FP.

The absolute distances of a point $P(x, y)$ on either curve from the directrices are the numerical values of

(35.13) $$x - \frac{a}{e}, \quad x + \frac{a}{e}.$$

The focal distances of P, being e times the absolute distances of P from the directrices, are the numerical values of

(35.14) $$ex - a, \quad ex + a.$$

For an ellipse the first of (35.13) is negative and the second positive, whatever be P, as is seen from Fig. 31. Hence

$$FP = a - ex, \quad F'P = a + ex,$$

from which it follows that

$$(35.15) \quad FP + F'P = 2\,a.$$

For a hyperbola both of the quantities (35.13) are positive when P is on the right-hand branch of the hyperbola (see Fig. 31′), and both are negative when P is on the left-hand branch. Hence in the first case (35.14) are the focal radii, and in the second case they are the negatives of these quantities. Accordingly we have

$$(35.15') \quad F'P - FP = \pm 2\,a,$$

the $+$ sign or the $-$ sign applying according as P is on the right-hand or left-hand branch.

Thus we have the theorems

[35.1] *The sum of the focal radii of any point of an ellipse is equal to a constant, the length of the major axis.*

[35.1′] *The numerical value of the difference of the focal radii of any point of a hyperbola is equal to a constant, the length of the transverse axis.*

These results make possible a continuous construction of the central conics, as distinguished from a point-by-point construction. For the construction of an ellipse two thumbtacks are fastened through a sheet of paper, and a loop of string is placed loosely around the tacks; the string is then drawn taut by the point of a pencil. As the pencil is made to move, the string being held taut, it describes an ellipse with the tacks at the foci. In fact, if the distance between the foci is denoted by $2\,c$ and the length of the loop by $2\,a + 2\,c$, the sum of the focal radii is $2\,a$. Since $c = ae$, an ellipse of given major axis $2\,a$ and given eccentricity e is described when the tacks are set at the distance $2\,ae$ apart and the loop is of length $2\,a(1 + e)$.

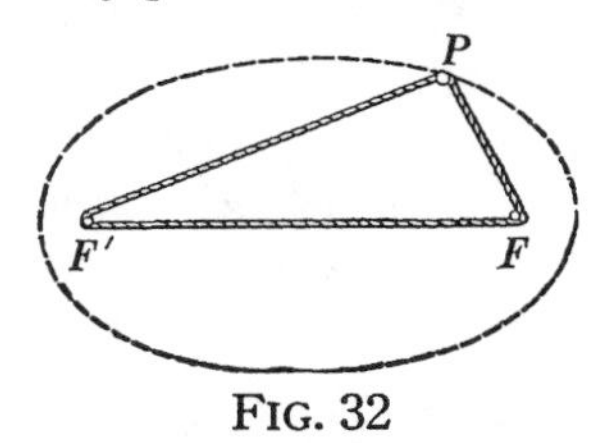

FIG. 32

For the construction of a hyperbola one end of a stick of length d is pivoted at a point F' on the paper, and at the other end of the stick is fastened one end of a string of length $d - 2a$; the other end of the string is then fastened by a thumbtack at a point F on the paper. If the string is held taut by pressing it with the point P of a pencil against the stick, the pencil describes a branch of a hyperbola as the stick rotates about the point F', since $F'P - FP = 2a$. Again, since $F'F = 2ae$, a hyperbola of given transverse axis and given eccentricity can be drawn by choosing the length of the string suitably and by setting the points F' and F at the proper distance apart.

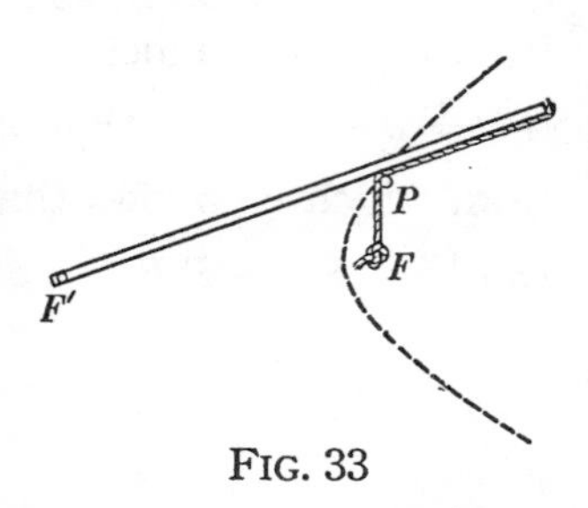

FIG. 33

As a result of the foregoing discussion we have that Theorems [**35.1**] and [**35.1'**] state characteristic properties of ellipses and hyperbolas, in the sense that these properties may be used to define these central conics. The same is true of the following two theorems, which are geometric statements of equations (35.10) and (35.10') respectively without regard to any coordinate system:

[**35.2**] *If d_1 and d_2 denote the distances of any point on an ellipse from its major and minor axes respectively, which axes have the respective lengths $2a$ and $2b$, then*

(35.16)
$$\frac{d_2^2}{a^2} + \frac{d_1^2}{b^2} = 1;$$

and any curve all of whose points are so related to two perpendicular line segments, of lengths $2a$ and $2b$ and each of which bisects the other, is an ellipse, whose eccentricity is given by $e^2 = (a^2 - b^2)/a^2$ when $a > b$.

[**35.3**] *If d_1 and d_2 denote the distances of any point on a hyperbola from the lines of its transverse and conjugate axes respectively, which axes have the respective lengths $2a$ and $2b$, then*

(35.17)
$$\frac{d_2^2}{a^2} - \frac{d_1^2}{b^2} = 1;$$

and any curve all of whose points are so related to two perpendicular line segments, of lengths $2a$ *and* $2b$ *and each of which bisects the other, is a hyperbola, whose eccentricity is given by* $e^2 = (a^2 + b^2)/a^2$.

For example, an equation of the ellipse whose major and minor axes are of lengths 6 and 4, and are on the lines $x + 1 = 0$ and $y - 2 = 0$ respectively, is

$$\frac{(y-2)^2}{9} + \frac{(x+1)^2}{4} = 1.$$

The conics were studied extensively by the Greeks from a purely geometric point of view, and many of their properties which now are derived by means of coordinate geometry were discovered by geometric reasoning (see § 38). Following the adoption of the Copernican theory of the planetary system, Kepler by laborious calculations from observational data showed that the orbits of the planets are ellipses with the sun at one of the foci. This enabled Newton to discover his law of gravitation, so that now the orbits of the planets are obtained readily from Newton's law by the use of coordinate geometry and the calculus. We have seen that for an ellipse each focus is at the distance ae from its center, and thus the eccentricity determines the departure of the focus from the center. For the earth e is about $1/60$, so that its orbit is almost circular. For the recently discovered planet Pluto e is about $1/4$, the semi-major axis of its orbit is nearly 40 times that of the earth, and its period of revolution is approximately 250 years. The paths of the comets are practically parabolas, some of them being ellipses of eccentricity almost equal to 1. In fact, for the celebrated Halley's comet $e = .98$, the semi-major axis is almost 18 times that of the earth's orbit, and its period is 75 years; its return to the earth's neighborhood has been recorded many times.

EXERCISES

1. Show that equation (35.10) in which $b > a$ is an equation of an ellipse with *semi-major axis* b and *semi-minor axis* a, that $(0, be)$ and $(0, -be)$ are the foci, and that $y = \pm b/e$ are the directrices, where $e^2 = (b^2 - a^2)/b^2$.

2. Find the vertices, center, foci, and directrices of the following:

a. $3x^2 + 4y^2 = 12$.
b. $9x^2 + 5y^2 = 45$.
c. $5x^2 - 4y^2 = 20$.
d. $9x^2 - 16y^2 = 12$.

3. Find an equation of an ellipse whose foci are $(-3, 0)$ and $(3, 0)$,
(*a*) when its minor axis is 8;
(*b*) when its major axis is twice its minor axis;
(*c*) when its eccentricity is 2/3.

4. Find an equation of a hyperbola with directrix $y = 1$, focus $(0, 3)$, and eccentricity 3/2.

5. Find an equation of a hyperbola whose transverse and conjugate axes, of lengths 4 and 6, are on the lines $x + 3 = 0$ and $y - 1 = 0$ respectively.

6. Find an equation of the ellipse whose major and minor axes, of lengths 8 and 6, are on the lines $3x - 4y + 1 = 0$ and $4x + 3y + 2 = 0$ respectively.

7. Find an equation of the locus of a point the sum of whose distances from the points $(c, 0)$ and $(-c, 0)$ is $2a$; also the locus when the difference of these distances is $2a$.

8. By definition the *latus rectum* of an ellipse (or a hyperbola) is the chord through a focus and perpendicular to the major (or transverse) axis; find its length in terms of a and b.

9. Show that, if the distance of a focus of an ellipse from the corresponding directrix is h, the semi-major axis is given by $a = he/(1 - e^2)$. What is the semi-transverse axis of the hyperbola with the same focus and directrix?

10. Show that the quantity $\frac{x_1^2}{a^2} + \frac{y_1^2}{b^2} - 1$ is positive or negative according as the point (x_1, y_1) lies outside or inside the ellipse (35.10). What is the similar theorem for the hyperbola?

11. In what sense is a circle an ellipse of eccentricity zero?

12. What is the eccentricity of a *rectangular hyperbola*, that is, one whose transverse and conjugate axes are of equal length? What is an equation of a rectangular hyperbola?

13. Prove that the distance of any point of a rectangular hyperbola (cf. Ex. 12) from the center is the mean proportional to its distances from the foci.

14. Find an equation of the ellipse with the point $(2, -1)$ as focus, the line $3x - 4y - 5 = 0$ as directrix, and eccentricity $1/2$; also find an equation of the hyperbola with this focus and directrix, and eccentricity 2. What is the major axis of this ellipse; the transverse axis of this hyperbola? (See Ex. 9.)

15. Given the equation $r = \dfrac{l}{1 - e\cos\theta}$ in polar coordinates; show by means of (29.5) and (29.6) that an equation of the locus in rectangular coordinates is

$$(1 - e^2)x^2 + y^2 = l^2 + 2\,lex,$$

and that this is an equation of a conic of eccentricity e with focus at the origin and the line $x + \dfrac{l}{e} = 0$ for directrix. Show that the latus rectum is of length $2\,l$.

16. Consider in space of three dimensions a circle of radius a with center at the origin, and lying in a plane which cuts the xy-plane in the x-axis and makes the angle θ with this plane; show that the orthogonal projection of the circle upon the xy-plane is an ellipse with the equation

$$\frac{x^2}{a^2} + \frac{y^2}{a^2\cos^2\theta} = 1.$$

Is the orthogonal projection of any circle upon a plane not parallel to the plane of the circle an ellipse?

36. Conjugate Diameters and Tangents of Central Conics

We turn now to the consideration of sets of parallel chords of a central conic and to the finding of equations of tangents.

The x-coordinates of the points in which the line

$$y = mx + h \tag{36.1}$$

meets the curves with equations (35.10) and (35.10′) are given by

(36.2)

$$x = \frac{-a^2mh \pm ab\sqrt{b^2 + a^2m^2 - h^2}}{a^2m^2 + b^2}.$$

(36.2′)

$$x = \frac{-a^2mh \pm ab\sqrt{b^2 - a^2m^2 + h^2}}{a^2m^2 - b^2}.$$

From these expressions and (36.1) we find as the coordinates x_0, y_0 of the mid-point of the line segment connecting the points of intersection

$$x_0, y_0 = \frac{-a^2mh,\ hb^2}{a^2m^2 + b^2}.$$

$$x_0, y_0 = \frac{-a^2mh,\ -hb^2}{a^2m^2 - b^2}.$$

As h takes on different values for which the points of intersection are real, m remaining fixed, that is, for different parallel chords, these expressions give the coordinates of the mid-point of each chord. By substitution we find that these are points on the respective lines

(36.3) $$y = -\frac{b^2}{a^2m}x.$$

(36.3′) $$y = \frac{b^2}{a^2m}x.$$

Each, passing through the origin, which is the center of the curve, and hence being a line through the center of a central conic, is called a *diameter*. If in each case we denote by m' the slope of the line (36.3) or (36.3′), we have

(36.4) $$mm' = -\frac{b^2}{a^2}.$$

(36.4′) $$mm' = \frac{b^2}{a^2}.$$

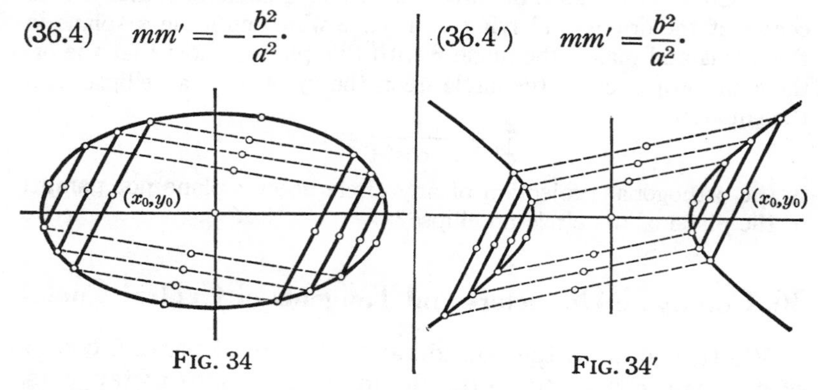

FIG. 34 FIG. 34′

Since these equations are symmetric in m and m', it follows that the mid-points of the chords parallel to the lines (36.3) and (36.3′) are on the line $y = mx$ in each case. Two lines through the center of an ellipse, or of a hyperbola, whose slopes are in the relation (36.4), or (36.4′), are called *conjugate diameters*. Hence we have

[36.1] *Either of two conjugate diameters of an ellipse, or of a hyperbola, is the locus of the mid-points of chords parallel to the other.*

Equations (36.3) and (36.3′) do not apply when $m = 0$, that is, when the chords are parallel to the major axis of an ellipse or the transverse axis of a hyperbola. But since the axes in each case are axes of symmetry, for each curve either axis is the locus of the mid-points of chords parallel to the other. Consequently the axes are said to be conjugate.

We leave to the reader the proof of the following statements, with the aid of (36.2) and (36.2′), calling his attention to the derivation in § 34 of similar results for the parabola:

The parallel lines

$$y = mx \pm \sqrt{a^2m^2 + b^2} \tag{36.5}$$

are tangent to the ellipse (35.10) at the points

$$\left(\pm \frac{a^2m}{\sqrt{a^2m^2+b^2}}, \pm \frac{b^2}{\sqrt{a^2m^2+b^2}}\right). \tag{36.6}$$

The parallel lines

$$y = mx \pm \sqrt{a^2m^2 - b^2} \tag{36.5'}$$

are tangent to the hyperbola (35.10′) at the points

$$\left(\pm \frac{a^2m}{\sqrt{a^2m^2-b^2}}, \mp \frac{b^2}{\sqrt{a^2m^2-b^2}}\right). \tag{36.6'}$$

On denoting by x_1, y_1 the coordinates of either point, the equation of the tangent at the point may be written in the form

$$\frac{xx_1}{a^2} + \frac{yy_1}{b^2} = 1. \tag{36.7}$$

$$\frac{xx_1}{a^2} - \frac{yy_1}{b^2} = 1. \tag{36.7'}$$

We now establish the following theorem, which shows that the central conics possess a property somewhat analogous to the property of the parabola stated in Theorem [**34.3**]:

[36.2] *The focal radii of any point P of an ellipse, or of a hyperbola, make equal angles with the tangent to the conic at P.*

We prove this theorem by using straightforward algebraic methods, which illustrate the power of coordinate geometry. In order to handle the ellipse and hyperbola at the same time, we use equation (35.6), that is,

$$\frac{x^2}{a^2} + \frac{y^2}{a^2(1-e^2)} = 1. \tag{36.8}$$

From equations (36.7), (36.7′) and (35.9), (35.9′) it follows that an equation of the tangent to either conic at the point $P_1(x_1, y_1)$ is

$$\frac{xx_1}{a^2} + \frac{yy_1}{a^2(1-e^2)} = 1.$$

Hence direction numbers of this line are (see page 30)

$$u_1 = \frac{y_1}{a^2(1-e^2)}, \qquad v_1 = -\frac{x_1}{a^2}.$$

Since the foci are $F'(-ae, 0)$ and $F(ae, 0)$, direction numbers of the focal radii are (see Theorem [6.1])

$$u_2 = x_1 \pm ae, \quad v_2 = y_1,$$

where the $+$ sign holds for $F'P_1$ and the $-$ sign for FP_1. We keep the sign $\pm$ in the expression for u_2, in order that the result later obtained shall apply to both $F'P_1$ and FP_1.

In order to apply formula (6.8), we first make the following calculations:

$$u_1u_2 + v_1v_2 = \frac{y_1}{a^2(1-e^2)}(x_1 \pm ae) - \frac{x_1y_1}{a^2} = \frac{y_1e}{a^2(1-e^2)}(\pm a + x_1e),$$

$$u_1^2 + v_1^2 = \frac{1}{a^4}\left(\frac{y_1^2}{(1-e^2)^2} + x_1^2\right),$$

$$u_2^2 + v_2^2 = (x_1 \pm ae)^2 + y_1^2.$$

Substituting x_1 and y_1 for x and y in (36.8), solving for y_1^2, and substituting in the last of the above equations, we have

$$\begin{aligned} u_2^2 + v_2^2 &= (x_1 \pm ae)^2 + a^2(1-e^2) - x_1^2(1-e^2) \\ &= a^2 \pm 2\,x_1ae + e^2x_1^2 = (\pm a + x_1e)^2. \end{aligned}$$

When these expressions are substituted in (6.8), where now ϕ is the angle between the tangent at P_1 and a focal radius, we obtain

$$\cos\phi = \frac{y_1e}{e_1e_2\sqrt{y_1^2 + x_1^2(1-e^2)^2}}.$$

Since the factor $(\pm a + x_1e)$, in which the $+$ sign refers to the focal radius $F'P_1$ and the $-$ sign to the focal radius FP_1, does not appear in the result, it follows that the angles which these focal radii make with the tangent to the conic at the point P_1 are equal, as was to be proved.

As a physical application of this theorem we have that a ray of light emanating from one focus of an ellipse and meeting the ellipse at a point P would be reflected by a mirror tangential to the ellipse at P into a ray which would pass through the other focus.

EXERCISES

1. Find equations of the tangents to the hyperbola

$$5x^2 - 4y^2 - 10 = 0$$

which are perpendicular to the line $x + 3y = 0$.

2. Find equations of the tangent and normal to $3x^2 - 4y^2 - 8 = 0$ at the point $(2, -1)$.

3. Find the extremities of the diameter of $x^2 + 2y^2 = 4$ which is conjugate to the diameter through the point $(1, 1)$.

4. Find the two tangents to the ellipse $5x^2 + 9y^2 = 45$ from the point $(2, -2)$.

5. Show that equation (36.7) when the point (x_1, y_1) is outside the ellipse is an equation of the line through the points of tangency of the two tangents to the ellipse from (x_1, y_1). This line is called the *polar* of the point (x_1, y_1) with respect to the ellipse (see the latter part of § 34).

6. Do the results of § 34, Ex. 13 hold also for an ellipse and for a hyperbola?

7. Show that any tangent to an ellipse meets the tangents at the vertices in points the product of whose distances from the major axis is equal to the square of the semi-minor axis.

8. Show that by a rotation of the axes (28.8) for $\theta = 90°$, equation (35.10′) is transformed into

$$\frac{x'^2}{b^2} - \frac{y'^2}{a^2} = -1,$$

and that in the $x'y'$-system the coordinates of the vertices are $(0, -a)'$, $(0, a)'$ and of the foci, $(0, -ae)'$, $(0, ae)'$; and equations of the directrices are $y' \pm \frac{a}{e} = 0$. Apply this result to the determination of the vertices, foci, and directrices of the hyperbola $3x^2 - 4y^2 = -12$.

9. Given an ellipse (35.10), for which $a > b$, and the circle with center at $O(0, 0)$ and radius a, through a point (x, y) on the ellipse draw the line perpendicular to the x-axis, and denote by P the point in which this line meets the circle, and by θ the angle which the radius OP makes with the x-axis. Show that $x = a\cos\theta$, $y = b\sin\theta$, which therefore are parametric equations of the ellipse.

10. Find the relation between the lengths of two conjugate diameters of an ellipse and also between those of a hyperbola. Show that when a hyperbola has a pair of conjugate diameters of equal length, it is rectangular.

11. Show that the product of the distances of the foci from any tangent to an ellipse is equal to the square of the semi-minor axis.

12. Show that the point in which a perpendicular from either focus of an ellipse upon any tangent meets the latter lies on a circle with center at the center of the ellipse and radius equal to the semi-major axis.

13. Show that the sum of the squares of the reciprocals of two perpendicular diameters of an ellipse is constant.

14. Show that

$$\frac{x^2}{a^2 - t} + \frac{y^2}{b^2 - t} = 1 \qquad (a > b),$$

for all values of t less than a^2 except b^2, defines a system of central conics, all of which have the same foci. This system is called a system of *confocal conics*.

15. Show that through each point in the plane not on either coordinate axis there pass two conics of a confocal system, one being an ellipse and the other a hyperbola, and that the tangents to these curves at the point are perpendicular to one another.

37. Similar Central Conics. The Asymptotes of a Hyperbola. Conjugate Hyperbolas

Consider in connection with equation (35.10) the equation

$$\frac{x^2}{a^2} + \frac{y^2}{b^2} = k, \tag{37.1}$$

where k is some constant. If k is positive, this equation may be written

$$\frac{x^2}{a'^2} + \frac{y^2}{b'^2} = 1,$$

where $a'^2 = ka^2$, $b'^2 = kb^2$, and consequently is an ellipse with the same center (the origin) as the ellipse (35.10). Furthermore if (x, y) is a point of (35.10), then $(\sqrt{k}\,x, \sqrt{k}\,y)$ is a point of the ellipse (37.1), and corresponding points lie on a line through the center; that is, the ellipses (37.1) for positive values of k

are similar (see § 32). From (35.11) and § 35, Ex. 1 it follows that the ellipses (37.1) for all positive values of k have the same eccentricity (see Theorem [**32.1**]).

When $k = 0$ in (37.1), we say that it is an equation of a *point ellipse*, since there is only one real solution of the equation, namely, $x = y = 0$. However, in this case $y = \pm\sqrt{-1}\frac{b}{a}x$, that is, we have two conjugate imaginary lines. When $k < 0$ in (37.1), we say that it is an equation of an *imaginary ellipse*, since there are no real solutions of the equation. Hence we have

[**37.1**] *Equation* (37.1) *with* $k > 0$ *is an equation of a family of similar ellipses having the same center and respective principal axes; equation* (37.1) *is an equation of a point ellipse when* $k = 0$, *and of a family of imaginary ellipses when* $k < 0$.

Similarly we consider in connection with (35.10′) the equation

$$\frac{x^2}{a^2} - \frac{y^2}{b^2} = k, \tag{37.2}$$

where k is some constant. Proceeding as in the case of equation (37.1), we have that for each positive value of k equation (37.2) is an equation of a hyperbola similar to the hyperbola with equation (35.10′) and having the same center and respective principal axes.

We consider next the case when $k = 0$, that is, the equation

$$\frac{x^2}{a^2} - \frac{y^2}{b^2} = 0. \tag{37.3}$$

When this equation is written in the form

$$\left(\frac{x}{a} - \frac{y}{b}\right)\left(\frac{x}{a} + \frac{y}{b}\right) = 0, \tag{37.4}$$

it is seen that the coordinates of a point on either of the lines

$$\frac{x}{a} - \frac{y}{b} = 0, \quad \frac{x}{a} + \frac{y}{b} = 0 \tag{37.5}$$

are a solution of (37.3), and, conversely, any solution of (37.3) is a solution of one of equations (37.5). Consequently (37.3) is an equation of the two lines (37.5) (see Ex. 7). We shall study the relation of these lines to the hyperbola (35.10′).

Since a and b are understood to be positive, the first of the lines (37.5), that is, $ay - bx = 0$, passes through the origin and lies in the first and third quadrants. In order to find the distance d from this line to a point (x_1, y_1) on the hyperbola in the first quadrant, we observe that it follows from (35.10′) that the y-coordinate of this point is given by $y_1 = \frac{b}{a}\sqrt{x_1^2 - a^2}$ in terms of x_1. Hence by Theorem [8.1] the distance d is given by

$$(37.6)\quad d = \frac{ay_1 - bx_1}{\sqrt{a^2+b^2}} = \frac{b(\sqrt{x_1^2 - a^2} - x_1)}{\sqrt{a^2+b^2}}$$

$$= \frac{b(\sqrt{x_1^2 - a^2} - x_1)(\sqrt{x_1^2 - a^2} + x_1)}{\sqrt{a^2+b^2}\,(\sqrt{x_1^2 - a^2} + x_1)}$$

$$= -\frac{ba^2}{\sqrt{a^2+b^2}} \cdot \frac{1}{\sqrt{x_1^2 - a^2} + x_1}.$$

As x_1 becomes larger and larger, that is, as the point (x_1, y_1) is taken farther out on the hyperbola, this distance becomes smaller and smaller, and we say that the hyperbola *approximates* the line for very large values of x. Since the center of the hyperbola, that is, the origin in this coordinate system, is a point of symmetry of the hyperbola, it follows that the same situation exists in the third quadrant for numerically large negative values of x. The reader can show that the second of the lines (37.5) bears a similar relation to the hyperbola in the second and fourth quadrants.

The lines (37.5) are called the *asymptotes* of the hyperbola; they pass through the center of the hyperbola and are equally inclined to its axes. For a rectangular hyperbola (see § 35, Ex. 12) the asymptotes are mutually perpendicular, and only for such a hyperbola.

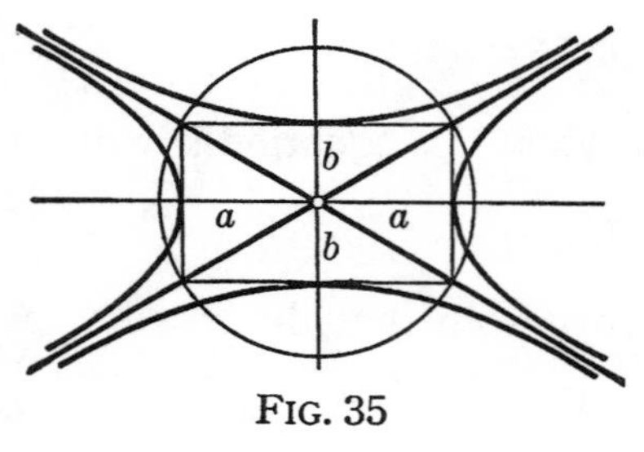

Fig. 35

The lines (37.5) by a similar argument are the asymptotes also of the hyperbola

$$(37.7)\quad \frac{x^2}{a^2} - \frac{y^2}{b^2} = -1.$$

The two hyperbolas (35.10′) and (37.7) are said to be *conjugate* to one another. They lie with respect to their common asymptotes as shown in Fig. 35, and the transverse axis of either hyperbola is the conjugate axis of the other. Equation (37.2) for any negative value of k is an equation of a hyperbola similar to (37.7). Hence we have

[37.2] *Equation* (37.2) *with* $k \neq 0$ *is an equation of two families of hyperbolas, the one family for* $k > 0$ *having the x-axis for transverse axis of each hyperbola, and the other family for* $k < 0$ *having the y-axis for transverse axis of each hyperbola; all of these hyperbolas have the same asymptotes, given by* (37.2) *for* $k = 0$.

The proof of the last part of this theorem is left to the reader as an exercise.

When the values b/a and $-b/a$ are substituted for m in (36.5′), we get the equations of the asymptotes. However, in the expressions (36.6′) for the coordinates of the points of contact the denominators become equal to zero for these values of m, and consequently this treatment of tangents cannot properly be applied to the asymptotes. Sometimes it is said that the asymptotes are tangent to a hyperbola at infinity, but we prefer the statement that the hyperbola approximates the asymptotes as distances from the center become very large.

EXERCISES

1. Find the center, vertices, foci, and directrices of the hyperbola conjugate to $3x^2 - 4y^2 = 12$.

2. Show that the circle in Fig. 35 meets the transverse axis of each hyperbola in its foci, and meets each hyperbola in points on the directrices of the other hyperbola.

3. Show by means of a transformation (28.8) that when the rectangular hyperbola $x^2 - y^2 = 2a^2$ is referred to its asymptotes as axes its equation is $xy = a^2$.

4. Show that the eccentricities e and e' of a hyperbola and its conjugate are in the relation $\frac{1}{e^2} + \frac{1}{e'^2} = 1$.

5. Find the points where the diameter of $x^2 - 2y^2 = 2$ conjugate to the diameter through $(-2, 1)$ meets the conjugate hyperbola.

6. Find equations of the tangents of slope $+1$ to the hyperbola $4x^2 - 3y^2 + 12 = 0$ and the points of contact; for what values of the slope are there tangents to this hyperbola?

7. Show that the graph of the equation

$$(a_1x + b_1y + c_1)(a_2x + b_2y + c_2) = 0$$

is the two lines $a_1x + b_1y + c_1 = 0$ and $a_2x + b_2y + c_2 = 0$.

8. Show that if the angle between the asymptotes of a hyperbola is denoted by 2α, the eccentricity of the hyperbola is $\sec \alpha$.

9. Show that the portion of an asymptote of a hyperbola included between the two directrices is equal to the length of the transverse axis.

10. Show that the distance of a focus of a hyperbola from either asymptote is equal to the semi-conjugate axis.

11. Find the bisectors of the angles between the lines joining any point on a rectangular hyperbola to its vertices, and determine their relation to the asymptotes.

12. Show that the point of contact of a tangent to a hyperbola is the mid-point of the segment of the tangent between the points in which it meets the conjugate hyperbola.

13. Show that if a line meets a hyperbola in the points P' and P'', and the asymptotes in R' and R'', the mid-points of the segments $P'P''$ and $R'R''$ coincide.

14. Show that the product of the distances of a point on a hyperbola from its asymptotes is constant.

15. Show that the lines joining either vertex of a hyperbola to the end points of its conjugate axis are parallel to the asymptotes.

16. Identify each of the loci defined by the equation

$$k_1 \frac{x^2}{a^2} + k_2 \frac{y^2}{b^2} = k_3$$

when each of the k's takes the values 0, $+1$, and -1; draw on one graph all these loci for $a = 1$, $b = 2$.

17. Show that two equations (37.2) for values of k equal numerically but of opposite sign are equations of conjugate hyperbolas.

18. Show that the ratio of the semi-axes for each family of hyperbolas (37.2) as $k > 0$ or $k < 0$ is the same for all members of the family. What relation does this ratio bear to the eccentricity of each hyperbola?

38. The Conics as Plane Sections of a Right Circular Cone

Consider a right circular cone with vertex O and a section MVN of the cone by a plane π, as in Fig. 36; V is the point of the curve MVN on the intersection $DVKL$ of π and the plane perpendicular to π through the axis OA; K is the intersection of π and OA, and C the point in which the bisector of the angle KVO meets OA. The point C is at the same distance r from the element OV of the cone and from the line of symmetry KV of the curve MVN; and the perpendicular from C upon KV is normal to the plane π, being in a plane perpendicular to π. With C as center and r as radius describe a sphere; denote by F the point where it is tangent to the plane of the section and by B the point where it is tangent to the line OV. Since the cone is right circular, this sphere is tangent to each element of the cone, and all the points of tangency are on a circle BE. DG is the line of intersection of the plane of the circle and the plane π; this line is perpendicular to BD.

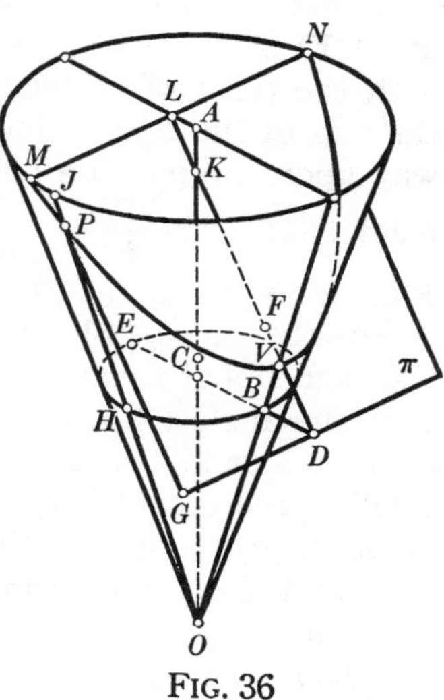

FIG. 36

Let P be any point of the curve in which the plane cuts the cone, and PHO the element of the cone through P, H being its point of tangency to the sphere. Since PF and PH are tangents to the sphere from an outside point, they are equal, as shown in solid geometry. Since all the elements of the cone make the same angle with the plane EHB, the line PH makes with this plane an angle equal to VBD. From P we draw PG perpendicular to DG, which being parallel to the line KFD makes the same angle with the plane of the circle EHB as the latter line, that is, the angle VDB. If we denote by Q (not shown in Fig. 36) the foot of the perpendicular from P on the plane of the circle, we have

$$PH = \frac{PQ}{\sin VBD}, \qquad PG = \frac{PQ}{\sin VDB}.$$

Hence we have

$$\frac{PF}{PG} = \frac{PH}{PG} = \frac{\sin VDB}{\sin VBD}. \tag{38.1}$$

The angle VBD is the complement of the angle AOB, and the angle VDB is the angle which the plane of the section makes with the plane of the circle EHB, which is a plane normal to the axis of the cone. Since these angles do not depend in any way upon the position of the point P on the curve, it follows from (38.1) that $\frac{PF}{PG} = \text{const.}$, and thus the curve is a conic, F being the focus and DG the directrix (see § 32).

For the curve to be an ellipse, the plane must intersect all the elements of the cone; that is, the angle VDB must be less than the angle VBD, in which case the ratio (38.1) is less than 1, as it should be. As the angle VDB is taken smaller and smaller, the eccentricity becomes smaller and approaches the value 0, in which case the plane is normal to the axis of the cone and the plane section is a circle. This is in agreement with (35.11), from which it follows that $e = 0$ when $a = b$.

When the angle VDB is equal to the angle VBD, that is, when the line DVL is parallel to the element OE, in which case the cutting plane is parallel to OE, the ratio (38.1) is equal to $+1$, and the section by the plane is a parabola.

When the angle VDB is greater than the angle VBD, in which case the ratio (38.1) is greater than $+1$, the plane intersects only some of the elements of the cone and the section is one branch of a hyperbola, the other branch being the section of the cone obtained by extending the elements through O. If we take a plane through O parallel to the cutting plane, it intersects the two cones in two elements; when these are projected orthogonally upon the cutting plane, the resulting lines are the asymptotes of the hyperbola, and their point of intersection, that is, the projection of O, is the center of the hyperbola.

When the conic section is an ellipse, there is a sphere above the cutting plane which is tangent to the plane at a point F' on the line VK and is tangent to the cone along a circle E', H', B', these points being on the elements through the respective points E, H, B. Since the planes of the two circles are parallel, the cutting plane meets the plane of the second circle in a line $D'G'$ parallel to DG, the points D' and G' being on the lines VK and GP respectively. As in the preceding case, $PH' = PF'$, being equal tangents from P to the second sphere. By the argument used above we show that F' is the other focus of the ellipse and $D'G'$ the corresponding directrix. Moreover, for any point P the sum of PF and PF' is equal to the length of the segment of an element of the cone between the planes of the two circles; and this length is equal to the length of the major axis of the ellipse, as is seen when P is taken at V or at V' (the point in which VK meets the cone again) (see Theorem [**35.1**]).

As previously remarked, when the conic section is a hyperbola one branch lies on each of the two cones which are a prolongation of one another through O. The focus within each branch is the point of contact of a sphere in each cone similar to the one first discussed in connection with Fig. 36. We leave it to the reader to show that the difference of the focal radii is equal to the segment of an element of the cones between the planes of the two circles of tangency of the spheres with the cones, the length of the segment being equal to the length of the transverse axis of the hyperbola (see Theorem [**35.1'**]).

39. Equations of Conics Whose Axes Are Parallel to the Coordinate Axes

It was remarked in § 32 that an equation of a curve is determined not only by the geometric character of the curve but also by its position relative to the coordinate axes. The equation

$$(1 - e^2)x^2 + y^2 - 2\,kx + k^2 = 0 \tag{39.1}$$

was derived in § 32 as an equation of a conic of eccentricity e, when the directrix is the y-axis, and the perpendicular through the focus upon the directrix the x-axis, the focus being denoted by $(k, 0)$. Also in §§ 33 and 35 it was shown that by a suitable translation of axes equation (39.1) was transformed into (33.3) for a parabola ($e = 1$), and into (35.10) and (35.10′) for an ellipse ($e < 1$) and a hyperbola ($e > 1$) respectively.

If the directrix of a conic is taken as the x-axis and the perpendicular through the focus upon the directrix as the y-axis, and the focus is denoted by $(0, k)$, an equation of the conic is

$$x^2 + (1 - e^2)y^2 - 2\,ky + k^2 = 0, \tag{39.2}$$

as the reader can verify directly, or can obtain from (39.1) by interchanging x and y.

Equations (39.1) and (39.2) are special cases of the equation

$$ax^2 + by^2 + 2\,fx + 2\,gy + c = 0, \tag{39.3}$$

in which not both a and b are zero. We consider this equation for the two cases when either a or b is equal to zero, and when neither is equal to zero.

***Case 1.* $a = 0$ or $b = 0$.** If $a = 0$ and $b \neq 0$, on dividing equation (39.3) by b and completing the square of the terms involving y, we obtain

$$\left(y + \frac{g}{b}\right)^2 + \frac{2\,f}{b}x + \frac{c}{b} - \frac{g^2}{b^2} = 0. \tag{39.4}$$

When $f \neq 0$, and this equation is written

$$\left(y + \frac{g}{b}\right)^2 = -\frac{2\,f}{b}\left(x + \frac{cb - g^2}{2\,bf}\right), \tag{39.5}$$

on applying the translation of axes defined by

$$x' = x + \frac{cb - g^2}{2\,bf}, \quad y' = y + \frac{g}{b}, \tag{39.6}$$

equation (39.5) is transformed into

$$y'^2 = 4\,a'x', \quad \text{where} \quad a' = -\frac{f}{2\,b}, \tag{39.7}$$

which is seen to be an equation of a parabola.

In the $x'y'$-system the vertex is at the point $x' = 0$, $y' = 0$, the focus at the point $(a', 0)$, and equations of the tangent at the vertex and of the directrix are $x' = 0$ and $x' + a' = 0$ respectively. If then one desires the coordinates of these points and equations of these lines in the xy-system, he has only to substitute from (39.6) and (39.7) the expressions for x', y', and a' in terms of x, y, and the coefficients of equation (39.3).

When $a = f = 0$, equation (39.3) is

$$by^2 + 2\,gy + c = 0.$$

As a quadratic in y it has two solutions, say

$$y = k_1, \quad y = k_2,$$

and consequently the locus consists of two lines parallel to the x-axis, distinct or coincident according as $k_1 \neq k_2$ or $k_1 = k_2$; imaginary if k_1 and k_2 are imaginary.

Accordingly we have

[39.1] *An equation $by^2 + 2\,fx + 2\,gy + c = 0$ is an equation of a parabola with axis parallel to the x-axis when $f \neq 0$; when $f = 0$, it is an equation of two lines parallel to the x-axis, which may be coincident if real, or which may be imaginary.*

Similar reasoning applied to equation (39.3) when $a \neq 0$ and $b = 0$ yields the theorem

[39.2] *An equation $ax^2 + 2\,fx + 2\,gy + c = 0$ is an equation of a parabola with axis parallel to the y-axis when $g \neq 0$; when $g = 0$, it is an equation of two lines parallel to the y-axis, which may be coincident if real, or which may be imaginary.*

Consider, for example, the equation

$$2\,y^2 + 3\,x - 4\,y + 4 = 0.$$

On dividing the equation by 2 and completing the square in y, we have

$$(y-1)^2 = -\tfrac{3}{2}\,x - 1 = -\tfrac{3}{2}(x + \tfrac{2}{3}).$$

Hence the locus is a parabola with axis $y - 1 = 0$ (that is, its axis is parallel to the x-axis) and with vertex $(-\,2/3, 1)$. Since the coefficient of x is negative, the parabola lies to the left of the vertex, and the focus is at the directed distance $-\,3/8$ from the vertex, that is, the focus is the point $(-\,25/24, 1)$; and an equation of the directrix is

$$x + \tfrac{2}{3} - \tfrac{3}{8} = x + \tfrac{7}{24} = 0.$$

In obtaining these results we have used the processes which led to Theorem [**39.1**]. The reader should adopt this method in the solution of any exercise.

Case 2. $a \neq 0$, $b \neq 0$. On completing the squares in the x's and in the y's in (39.3), we obtain

$$a\left(x+\frac{f}{a}\right)^2+b\left(y+\frac{g}{b}\right)^2=k, \tag{39.8}$$

where by definition

$$k=\frac{f^2}{a}+\frac{g^2}{b}-c. \tag{39.9}$$

If then we effect the translation of the axes defined by

$$x'=x+\frac{f}{a}, \quad y'=y+\frac{g}{b}, \tag{39.10}$$

in the new coordinates (39.8) becomes

$$\frac{x'^2}{\frac{1}{a}}+\frac{y'^2}{\frac{1}{b}}=k. \tag{39.11}$$

When we compare this equation with equations (37.1) and (37.2), in consequence of Theorems [**37.1**] and [**37.2**] we have

[**39.3**] *An equation*

$$ax^2+by^2+2fx+2gy+c=0$$

in which a and b have the same sign is an equation of an ellipse (or circle when $a=b$), real or imaginary according as k, defined by (39.9), *has the same sign as a and b or a different sign from a and b, and of a point ellipse when $k=0$; when a and b have different signs, it is an equation of a hyperbola or intersecting lines according as $k \neq 0$ or $k=0$; for a real ellipse or a hyperbola the principal axes are on the lines*

$$y+\frac{g}{b}=0, \quad x+\frac{f}{a}=0. \tag{39.12}$$

The last part of this theorem follows from (39.10).

Consider, for example, the equation

$$2x^2-3y^2-4x-12y-6=0.$$

Completing the squares in the x's and in the y's, we have

$$2(x-1)^2-3(y+2)^2=-4,$$

or

$$\frac{(x-1)^2}{2}-\frac{(y+2)^2}{\frac{4}{3}}=-1.$$

By Theorems [37.2] and [39.3] the locus is a hyperbola whose semi-transverse axis $2/\sqrt{3}$ is on the line $x - 1 = 0$ and semi-conjugate axis $\sqrt{2}$ is on the line $y + 2 = 0$. Its center is the point $(1, -2)$, and equations of the asymptotes are

$$\frac{x-1}{\sqrt{2}} - \frac{\sqrt{3}(y+2)}{2} = 0, \quad \frac{x-1}{\sqrt{2}} + \frac{\sqrt{3}(y+2)}{2} = 0.$$

For a hyperbola (37.7) the equation analogous to (35.11′) is $e^2 = (a^2 + b^2)/b^2$; hence for the hyperbola under consideration $e = \sqrt{5/2}$. Since the foci are on the line $x - 1 = 0$, and above and below the center at the distance be, they are the points $(1, -2 \pm \sqrt{10/3})$. Equations of the directrices are

$$y' \pm \frac{b}{e} = y + 2 \pm 2\sqrt{\tfrac{2}{15}} = 0.$$

EXERCISES

1. Draw the graphs of the following equations after finding the vertex, axis, and focus of each curve:

$$4y^2 - 32x + 4y - 63 = 0, \quad 3x^2 + 6x + 3y + 4 = 0.$$

2. Find equations of two ellipses with center at $(-2, 4)$, and principal axes parallel to the coordinate axes, the semi-axes being 4 and 3.

3. Find an equation of a hyperbola with center at $(2, -1)$, one end of the transverse axis at $(5, -1)$, and one end of the conjugate axis at $(2, -4)$. What is an equation of the conjugate hyperbola?

4. Find an equation of the parabola whose axis is parallel to the y-axis and which passes through the points $(0, 0)$, $(-1, 2)$, and $(2, 2)$.

5. Draw the graphs of the following equations, after finding the center and principal axes of each curve:

$$4x^2 + y^2 + 4x - 6y + 9 = 0,$$
$$9x^2 - 4y^2 - 18x - 8y - 31 = 0,$$
$$4x^2 - y^2 + 2x - 3y - 2 = 0.$$

6. Determine h so that the line $y = 2x + h$ shall be tangent to the first conic in Ex. 5.

7. Show that the eccentricity of an ellipse (39.3) is $\sqrt{\frac{b-a}{b}}$ or $\sqrt{\frac{a-b}{a}}$ according as the major axis of the ellipse is parallel to the x-axis or to the y-axis (see equation (39.11)).

8. Show that an equation of a parabola with axis parallel to the x-axis and which passes through the noncollinear points (x_1, y_1), (x_2, y_2), and (x_3, y_3) is

$$\begin{vmatrix} y^2 & x & y & 1 \\ y_1{}^2 & x_1 & y_1 & 1 \\ y_2{}^2 & x_2 & y_2 & 1 \\ y_3{}^2 & x_3 & y_3 & 1 \end{vmatrix} = 0.$$

Discuss this equation when the three points are collinear (see § 4, Ex. 9). What is an equation of the parabola through these points when its axis is parallel to the y-axis?

9. How many points are necessary to determine an ellipse or a hyperbola whose principal axes are parallel to the coordinate axes? Using Theorem [**27.2**], derive for this case an equation analogous to that of Ex. 8.

40. The General Equation of the Second Degree. Invariants

The most general equation of the second degree in x and y is of the form

$$(40.1) \qquad ax^2 + 2\,hxy + by^2 + 2\,fx + 2\,gy + c = 0.$$

In this section we shall show that any such equation is an equation of a conic, including the case of two lines, called a *degenerate conic*. In consequence of the results of § 39 it follows that all we have to do is to show that by a suitable rotation of the axes equation (40.1) is transformed into an equation in x' and y' in which there is no term in $x'y'$.

If now we apply the transformation (28.8) to equation (40.1), we obtain

$$(40.2) \quad a'x'^2 + 2\,h'x'y' + 2\,b'y'^2 + 2\,f'x' + 2\,g'y' + c' = 0,$$

where the coefficients are given by

$$(40.3) \quad \begin{aligned} a' &= a\cos^2\theta + 2\,h\sin\theta\cos\theta + b\sin^2\theta, \\ h' &= (b-a)\sin\theta\cos\theta + h(\cos^2\theta - \sin^2\theta), \\ b' &= a\sin^2\theta - 2\,h\sin\theta\cos\theta + b\cos^2\theta, \\ f' &= f\cos\theta + g\sin\theta, \quad g' = -f\sin\theta + g\cos\theta, \quad c' = c. \end{aligned}$$

From the above expression for h' it is seen that for the coefficient of $x'y'$ in (40.2) to be zero θ must be such that

$$(40.4) \qquad (b-a)\sin\theta\cos\theta + h(\cos^2\theta - \sin^2\theta) = 0.$$

When $b = a$, this equation is satisfied by $\cos\theta = \sin\theta$; that is, $\theta = 45°$. When $b \neq a$, if equation (40.4) is divided by $\cos^2\theta$, the resulting equation may be written

$$(40.5) \qquad h\tan^2\theta + (a-b)\tan\theta - h = 0,$$

from which we have, by means of the quadratic formula,

$$(40.6) \qquad \tan\theta = \frac{b - a \pm \sqrt{(b-a)^2 + 4h^2}}{2h}.$$

If for the moment we denote by θ_1 and θ_2 the values of θ corresponding to the signs $+$ and $-$ respectively before the radical, we find that

$$\tan\theta_1 \tan\theta_2 = -1.$$

This means that the x'-axis for the angle θ_1 and the one for θ_2 are perpendicular to one another (see Theorem [7.3]). However, we are interested in finding a rotation of the axes so that h' shall be zero, and consequently either solution (40.6) yields the desired result. Since such a transformation is always possible, we have shown that any equation of the second degree in x and y is an equation of a conic.

For example, if it is required to transform the equation

$$5x^2 - 4xy + 2y^2 + 2x - 2y - 1 = 0$$

so that in the new coordinates there is no term in $x'y'$, in this case equation (40.5) is $2\tan^2\theta - 3\tan\theta - 2 = 0$. Solving this equation by means of the quadratic formula, we obtain $\tan\theta = 2$ or $-1/2$. Taking the first root, we have $\sin\theta = 2/\sqrt{5}$, $\cos\theta = 1/\sqrt{5}$, so that the desired transformation (28.8) is

$$x = \frac{1}{\sqrt{5}}(x' - 2y'), \qquad y = \frac{1}{\sqrt{5}}(2x' + y').$$

Substituting these expressions for x and y in the above equation and collecting terms, we obtain as the new equation

$$x'^2 + 6y'^2 - \frac{2}{\sqrt{5}}x' - \frac{6}{\sqrt{5}}y' - 1 = 0,$$

which, on completing the squares in x' and y', assumes the form

$$\left(x' - \frac{1}{\sqrt{5}}\right)^2 + 6\left(y' - \frac{1}{2\sqrt{5}}\right)^2 = \frac{3}{2}.$$

Consequently the curve is an ellipse, and with regard to the coordinate system x', y' the center of the ellipse is at the point $(1/\sqrt{5}, 1/2\sqrt{5})'$, its semi-major axis is $\sqrt{3/2}$, and its semi-minor axis $1/2$. The major axis has the slope 2 in the xy-system.

From (40.3) we derive other results of importance. On adding the expressions for a' and b', and making use of the fundamental identity

$$\sin^2\theta + \cos^2\theta = 1, \tag{40.7}$$

we obtain

$$a' + b' = a + b. \tag{40.8}$$

It is something of an exercise to calculate $a'b' - h'^2$ from (40.3), but the reader will feel repaid when he finds that with the aid of (40.7) the result is reducible to

$$a'b' - h'^2 = ab - h^2. \tag{40.9}$$

Thus we have found that when the coordinates in the general expression of the second degree which is the left-hand member of equation (40.1) are subjected to any transformation of the form (28.8), the expressions $a + b$ and $ab - h^2$ are equal to the same expressions in the coefficients of the *transform* (40.2), that is, the equation into which equation (40.1) is transformed. In this sense we say that $a + b$ and $ab - h^2$ are *invariants* under the transformation. Invariants are of fundamental importance in applications of transformations of coordinates, as we shall see in what follows.

Returning to the general equation (40.1), we consider the following three possible cases:

Case 1. $ab - h^2 = 0$. From (40.9) it follows that in a coordinate system for which $h' = 0$ either a' or b' must be zero, but not both, otherwise (40.2) is of the first degree, and consequently (40.1) is of the first degree. This latter statement follows also from the fact that if $a' = b' = 0$, then from (40.8) we have $b = -a$, and $ab - h^2 = -(a^2 + h^2) = 0$, which can hold for real values of a and h only when both are zero. When $h' = 0$, and either $a' = 0$ or $b' = 0$, we have from Theorems [**39.1**] and [**39.2**] that the curve is a parabola or two parallel lines. Conversely, when the curve is a parabola or two parallel lines, $ab - h^2 = 0$, as follows from (40.9). This is an example of the fact that when an invariant is equal to zero for one coordinate system, it is equal to zero for every coordinate system.

Case 2. $\boldsymbol{ab - h^2 > 0}$. When $h' = 0$, it follows from (40.9) that a' and b' are both positive or both negative, and in consequence of Theorem [**39.3**] it follows that (40.1) is an equation of a real or imaginary ellipse or a point ellipse.

Case 3. $\boldsymbol{ab - h^2 < 0}$. When $h' = 0$, it follows from (40.9) that a' and b' differ in sign, and in consequence of Theorem [**39.3**] it follows that (40.1) is an equation of a hyperbola or two intersecting lines.

Gathering together these results, we have the following theorem:

[**40.1**] *Equation* (40.1) *is an equation of a parabola or two parallel or coincident lines when* $ab - h^2 = 0$; *of a real or imaginary ellipse or a point ellipse when* $ab - h^2 > 0$; *of a hyperbola or two intersecting real lines when* $ab - h^2 < 0$.

This result may be given another form. We consider first the case $ab - h^2 = 0$, from which it follows that a and b have the same sign; they can be taken as positive; for, if they are not positive, by changing the sign of every term in (40.1) we make them positive. Consider now the terms of the second degree in (40.1), that is, $ax^2 + 2\,hxy + by^2$. According as h is positive or negative, when h is replaced by $\sqrt{ab}$ or $-\sqrt{ab}$, the above expression may be written $(\sqrt{a}\,x \pm \sqrt{b}\,y)^2$; that is, the terms of the second degree form a perfect square.

If $ab - h^2 \neq 0$ and $b \neq 0$, the terms of the second degree are equal to

$$\frac{1}{b}(by + hx + \sqrt{h^2 - ab}\,x)(by + hx - \sqrt{h^2 - ab}\,x),$$

as the reader can verify by multiplying these terms together. These factors are real or conjugate imaginary according as $ab - h^2 < 0$ or > 0. If $b = 0$, in which case $ab - h^2 < 0$, the factors of the terms of the second degree are x and $ax + 2\,hy$, both real and distinct. From this result and Theorem [**40.1**] we have

[**40.2**] *An equation* (40.1) *is an equation of a parabola or of two parallel or coincident lines when the terms of the second degree are a perfect square; it is an equation of an ellipse, which may be real, imaginary, or a point ellipse, when the factors of the terms of the second degree are conjugate imaginary; it is an equation of a hyperbola or two real intersecting lines when the factors of the terms of the second degree are real and distinct.*

We consider now the equation

$$(a_1x + b_1y + c_1)(a_2x + b_2y + c_2) = k, \tag{40.10}$$

where all the coefficients are real. In consequence of Theorem [**40.2**], if the factors $a_1x + b_1y$ and $a_2x + b_2y$ of the terms of the second degree are different, (40.10) is a hyperbola or two intersecting lines according as $k \neq 0$ or $k = 0$. When $k \neq 0$, and the axes are rotated so that there is no term in $x'y'$, and an appropriate translation of the axes is made, if necessary, each of the factors in (40.10) is transformed into a factor homogeneous of the first degree, and these are such that the terms of the second degree consist of the difference of a multiple of x'^2 and a multiple of y'^2. Hence we have

$$(a_1'x' + b_1'y')(a_1'x' - b_1'y') = k. \tag{40.11}$$

From this result, the discussion leading up to Theorem [**39.3**], and the definition of asymptotes in § 37 it follows that the lines whose equations are obtained on equating to zero the factors in the left-hand member of (40.11), and consequently those in (40.10), are the asymptotes of the hyperbola. Hence we have

[**40.3**] *An equation* (40.10) *for* $k \neq 0$, *such that the lines*

$$a_1x + b_1y + c_1 = 0, \quad a_2x + b_2y + c_2 = 0 \tag{40.12}$$

intersect, is an equation of a hyperbola for which these lines are the asymptotes; and (40.10) *for a suitable value of* k *is an equation of any hyperbola having the lines* (40.12) *for asymptotes.*

The proof of the second part of this theorem is left to the reader (see Ex. 4).

We consider next the case when the lines (40.12) are parallel or coincident, in which case equation (40.10) may be written

$$(40.13) \qquad (ax + by + c_1)(ax + by + c_2) = k.$$

If we effect a transformation of coordinates for which

$$ax + by + c_1 = 0$$

is the x'-axis, this equation becomes

$$y'(y' + c_2') = \frac{k}{a^2 + b^2},$$

where $c_2' = 0$ if $c_2 = c_1$, and $c_2' \neq 0$ if $c_2 \neq c_1$. In either case this is an equation of two parallel or coincident lines, depending upon the values of c_2', k, a, and b. Hence we have

[40.4] *An equation* $(ax + by + c_1)(ax + by + c_2) = k$ *is an equation of two real or imaginary, parallel or coincident lines.*

As a consequence of this result and Theorem [**40.2**] we have

[40.5] *When the terms of the second degree in an equation* (40.1) *are a perfect square, the locus is two parallel, or coincident, lines or a parabola according as the equation can or cannot be put in the form* (40.13).

Consider the equation

$$(40.14) \qquad x^2 + xy - 2y^2 - 2x + 5y - 2 = 0.$$

The terms of the second degree have the real factors $x - y$ and $x + 2y$, and consequently we seek constants d and e so that the above equation shall assume the form

$$(40.15) \qquad (x - y + d)(x + 2y + e) = k.$$

When the two expressions in parentheses are multiplied together, we get

$$x^2 + xy - 2y^2 + (d + e)x + (2d - e)y + de - k = 0.$$

Comparing this equation with (40.14), we see that we must have

$$d + e = -2, \qquad 2d - e = 5, \qquad de - k = -2;$$

from the first two of these equations it follows that $d = 1$, $e = -3$, so that for $de - k$ to be equal to -2, k must be -1. Since $k \neq 0$, (40.14) is an equation of a hyperbola and

$$x - y + 1 = 0 \quad \text{and} \quad x + 2y - 3 = 0$$

are its asymptotes. From these results it follows that the equation obtained from (40.14), on replacing the constant term -2 by -3, is an equation of two straight lines, since in this case k in (40.15) is equal to zero.

In analyzing a general equation of the second degree for which $ab - h^2 < 0$, it is advisable to use the above process, which yields the asymptotes when the locus is a hyperbola, and equations of the lines when the locus is two intersecting lines.

EXERCISES

1. Determine the type of conic defined by each of the following equations either by effecting a rotation of the axes or by using the theorems of this section:

a. $5x^2 - 4xy + 8y^2 + 18x - 36y + 9 = 0.$
b. $x^2 - 4xy + 4y^2 + 5y - 9 = 0.$
c. $2x^2 + 3xy - 2y^2 - 11x - 2y + 12 = 0.$
d. $x^2 + 2xy + y^2 - 2x - 2y - 3 = 0.$
e. $x^2 - 4xy - 2y^2 - 2x + 7y - 3 = 0.$

2. Find the asymptotes of the hyperbola

$$2x^2 - 3xy - 2y^2 + 3x - y + 8 = 0,$$

and derive therefrom equations of the principal axes of the hyperbola.

3. Find an equation of all hyperbolas which have the coordinate axes for asymptotes.

4. Show that a hyperbola is completely determined by its asymptotes and a point of the hyperbola, and apply this principle to find an equation of a hyperbola whose asymptotes are $2x - 3y + 1 = 0$, $x + y - 3 = 0$ and which passes through the point $(1, -2)$.

5. Show that the centers of all hyperbolas whose asymptotes are parallel to the coordinate axes and which pass through the points $(2, 5)$ and $(3, 2)$ lie on the line $3x - y - 4 = 0$; also find an equation of the hyperbola of this set which passes through the point $(-2, 3)$.

6. Find the eccentricity of a hyperbola with equation (40.10), making use of Ex. 8 of § 37.

7. Verify the following statement: An equation of a conic involves a term in xy, if and only if the directrices (or directrix) are not parallel to one of the coordinate axes.

41. The Determination of a Conic from Its Equation in General Form

Having shown in § 40 that the locus of any equation of the second degree in x and y is a conic, or a degenerate conic, we show in this section how one may determine whether a conic is degenerate or not directly from the coefficients of its equation without effecting any transformation of coordinates. The method used to establish these criteria gives at the same time a ready means of drawing the graph of such an equation.

We consider then the equation

$$ax^2 + 2\,hxy + by^2 + 2\,fx + 2\,gy + c = 0, \tag{41.1}$$

and discuss first the case when $b \neq 0$, writing the equation in the form

$$by^2 + 2(hx + g)y + (ax^2 + 2\,fx + c) = 0. \tag{41.2}$$

Considering this as a quadratic in y with x entering into the coefficients, and solving for y by means of the quadratic formula, we have

$$y = -\frac{hx + g}{b} \pm \frac{1}{b}\sqrt{A}, \tag{41.3}$$

where A is defined by

$$\begin{aligned} A &= (hx + g)^2 - b(ax^2 + 2\,fx + c) \\ &= (h^2 - ab)x^2 + 2(hg - bf)x + g^2 - bc. \end{aligned} \tag{41.4}$$

For any value of x for which A is positive, the two values of y given by (41.3) are the y-coordinates of two points on the curve having the given value of x for x-coordinate. For a value of x, if any, for which $A = 0$, equation (41.3) reduces to

$$y = -\frac{hx + g}{b}, \tag{41.5}$$

that is, the point having for coordinates a value of x for which $A = 0$ and the value of y given by (41.5) is a point of intersection of the curve and the line

$$hx + by + g = 0. \tag{41.6}$$

We may interpret (41.3) as follows: Draw, as in Fig. 37, the graph of the line (41.6); at any point on the line whose x-coordinate, say x_1, is such that A is positive, add and subtract from the y-coordinate of the point on the line the quantity $\frac{1}{b}\sqrt{A_1}$, where A_1 is the value of A when x has been replaced by x_1; the two values thus obtained are the y-coordinates of the two points on the curve for which the x-coordinate is x_1. In other words, these are the coordinates of the two points in which the curve is met by the line $x - x_1 = 0$. If x_0 is such that for this value of x the quantity A is equal to zero, the two points of intersection coincide in the point on the line for which $y = -\dfrac{hx + g}{b}$; hence the line $x - x_0 = 0$ is tangent to the curve at this point (see § 12, after equation (12.8)) if the curve is not a degenerate conic. Fig. 37 is only suggestive; the position of the line $hx + by + g = 0$ and the position and shape of the curve depend upon the values of the coefficients in (41.1).

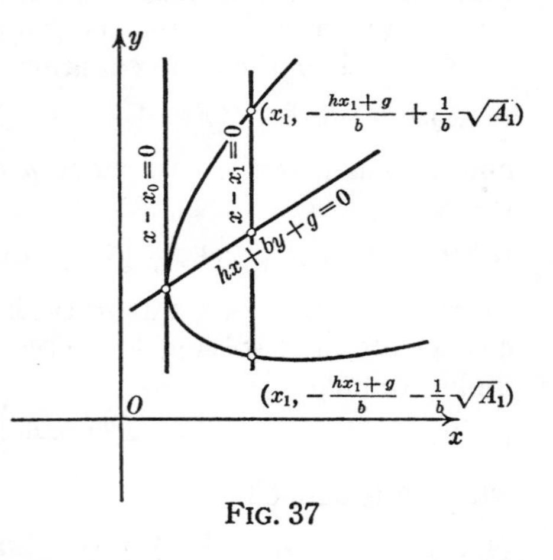

FIG. 37

From (41.3) it follows that (41.1) is an equation of two lines, if and only if A is a constant or the square of an expression of the first degree in x. In order that A shall be a constant, we must have

$$ab - h^2 = 0, \quad hg - bf = 0, \tag{41.7}$$

in which case (41.3) is

$$y = -\frac{hx+g}{b} \pm \frac{1}{b}\sqrt{g^2 - bc}, \tag{41.8}$$

which are equations of two parallel lines, real if $g^2 - bc > 0$, imaginary if $g^2 - bc < 0$, and of real and coincident lines if $g^2 - bc = 0$.

We consider next the case when $ab - h^2 \neq 0$ and A is a perfect square. The condition that A be a perfect square is

$$(hg - bf)^2 - (h^2 - ab)(g^2 - bc) = 0. \tag{41.9}$$

The expression on the left reduces to

$$-b(abc - ag^2 - bf^2 - ch^2 + 2fgh),$$

which is equal to $-bD$, where D is defined by

$$D = \begin{vmatrix} a & h & f \\ h & b & g \\ f & g & c \end{vmatrix}. \tag{41.10}$$

Since by hypothesis $b \neq 0$, we have that (41.9) is equivalent to $D = 0$. When this condition is satisfied, equations (41.3) are equivalent to

$$by + hx + g \mp (\sqrt{h^2 - ab}\, x + e\sqrt{g^2 - bc}) = 0, \tag{41.11}$$

since the square of the expression in parenthesis is equal to A in consequence of (41.9), e being $+1$ or -1 according as $hg - bf > 0$ or < 0.

We consider now the above results for the various possibilities as $ab - h^2$ is zero, positive, and negative, and also when $b = 0$, in which case equation (41.1) is

$$ax^2 + 2\,hxy + 2\,fx + 2\,gy + c = 0. \tag{41.12}$$

As a result of this analysis the reader will be able to determine completely the character of the locus of any equation of the second degree from the values of the coefficients of the equation. It is suggested that, as he proceeds, he make a table of the results of the analysis.

***Case 1.* $ab - h^2 = 0$.** When $b \neq 0$ and $D = 0$, we have (41.7), as follows from (41.9), and consequently the locus is two parallel or coincident lines (41.8). When $b = 0$, then $h = 0$, and (41.12) is an equation of two parallel or coincident lines, if and only if $g = 0$, as follows from Theorem [**39.2**]. When $b = h = g = 0$, we have $D = 0$. Accordingly we have

[**41.1**] *When* $ab - h^2 = 0$, *the locus of equation* (41.1) *is a parabola or two lines according as* $D \neq 0$ *or* $D = 0$; *when* $D = 0$ *and* $b \neq 0$, *the lines are parallel and real or imaginary according as* $g^2 - bc > 0$ *or* < 0, *and real and coincident if* $g^2 - bc = 0$; *when* $D = 0$ *and* $b = h = 0$, *the lines are parallel and real or imaginary according as* $f^2 - ac > 0$ *or* < 0, *and real and coincident when* $f^2 - ac = 0$.

The part of this theorem for $D = 0$, $b \neq 0$ follows from (41.8). When $b = h = 0$ it follows from $D = 0$ that $g = 0$. Hence we have

$$x = \frac{-f \pm \sqrt{f^2 - ac}}{a},$$

which is obtained on solving (41.12) in this case.

Case 2. $ab - h^2 > 0$. Since this case does not arise when $b = 0$, we have that when $D = 0$, equation (41.1) is equivalent to two conjugate imaginary equations of the first degree (41.11), whose common solution is real, and consequently (41.1) is an equation of a point ellipse, which is a degenerate conic. When $D \neq 0$, (41.1) is an equation of a real or imaginary ellipse. In order to distinguish these two types of an ellipse, we observe that for the ellipse to be real there must be real values of x for which A as defined by (41.4) is equal to zero, as follows when one considers Fig. 37 for the case of an ellipse. The condition for this to be so is that the left-hand member of (41.9) shall be positive, that is, $-bD > 0$. Hence we have

[**41.2**] *When* $ab - h^2 > 0$, *the locus of equation* (41.1) *is a point ellipse if* $D = 0$; *if* $D \neq 0$, *it is a real or imaginary ellipse according as* $-bD > 0$ *or* < 0.

Case 3. $ab - h^2 < 0$. When $b \neq 0$, it follows from (41.11) and Theorem [**40.1**] that (41.1) is an equation of a hyperbola or two intersecting lines according as $D \neq 0$ or $D = 0$. When $b = 0$, the factors of the second-degree terms in (41.12) are x

and $ax + 2\,hy$. If then (41.12) is to be an equation of two lines, there must exist numbers d and e such that

$$(x + d)(ax + 2\,hy + e)$$

is equal to the left-hand member of (41.12). Multiplying these expressions together and equating the coefficients of x and y and the constant terms of the two expressions, we have

$$e + ad = 2\,f, \qquad hd = g, \qquad ed = c.$$

Solving the first two equations for e and d and substituting in the third, we have that the following condition must be satisfied:

$$ag^2 + ch^2 - 2\,fgh = 0,$$

which, since $b = 0$, is in fact $D = 0$. Hence we have

[41.3] *When* $ab - h^2 < 0$, *the locus of equation* (41.1) *is a hyperbola or two intersecting lines according as* $D \neq 0$ *or* $D = 0$.

From the foregoing theorems we have

[41.4] *An equation* (41.1) *is an equation of a degenerate conic, if and only if the determinant* D *is equal to zero.*

The algebraic equivalent of this theorem is the following:

[41.5] *A quadratic expression*

$$ax^2 + 2\,hxy + by^2 + 2\,fx + 2\,gy + c$$

is equal to the product of two factors of the first degree, if and only if $D = 0$.

It follows from this theorem that a *quadratic form*

$$ax^2 + by^2 + cz^2 + 2\,hxy + 2\,fxz + 2\,gyz$$

is the product of two linear homogeneous factors in x, y, and z, if and only if $D = 0$. (See page 267.)

When by means of Theorems [**41.4**] and [**41.2**] the reader finds that the locus of a given equation is a degenerate conic which is not a point ellipse, he should reduce the equation to the product of two factors of the first degree in x and y, after the manner of the exercise worked toward the end of § 40, and interpret the result geometrically.

EXERCISES

1. Using the method of this section, draw the graphs of the equations of Ex. 1, § 40.

2. For what value of c is

$$x^2 - 3\,xy + 2\,y^2 + x - y + c = 0$$

an equation of two intersecting lines? Find equations of the lines.

3. For what values of k is $D = 0$ for equations (40.10) and (40.13)?

4. Show that when (41.1) is an equation of a real ellipse, the values of x for which A, defined by (41.4), is positive lie between the roots of the equation $A = 0$; when (41.1) is an equation of a hyperbola, the values of x for which A is positive are less than the smaller of the two roots and greater than the larger of the roots of $A = 0$ if the roots are real, that is, if $-bD > 0$. Discuss the case when $-bD < 0$.

5. Show that the line (41.6) is the locus of the mid-points of a set of parallel chords of the conic; and that when the conic is a parabola it is a line parallel to the axis of the parabola (see Theorem [**34.1**]).

6. Show that when (41.1) with $b \neq 0$ is an equation of a central conic, the x-coordinate of the center is one half the sum of the roots of the equation $A = 0$, where A is defined by (41.4), and that the coordinates x_0, y_0 of the center are given by

$$x_0 = \frac{bf - hg}{h^2 - ab}, \qquad y_0 = \frac{ag - hf}{h^2 - ab}.$$

Discuss the case when $b = 0$.

7. Assume that $a \neq 0$ in equation (41.1); consider the equation as a quadratic in x with y entering in the coefficients, and discuss the solution for x in a manner similar to that developed in this section.

8. Show that, when an equation of a hyperbola is written in the form (40.10), equations of its principal axes are (see (10.10))

$$\frac{a_1x + b_1y + c_1}{e_1\sqrt{a_1{}^2 + b_1{}^2}} = \pm\frac{a_2x + b_2y + c_2}{e_2\sqrt{a_2{}^2 + b_2{}^2}}.$$

Apply this result to find the principal axes of the hyperbola (40.14).

9. Show that when $ab - h^2 \neq 0$ for an equation (41.1) there exists a number $\bar{c}$ such that

$$ax^2 + 2\,hxy + by^2 + 2\,fx + 2\,gy + \bar{c}$$

is the product of two factors of the first degree.

10. Show that in consequence of Ex. 9 and (41.11) equation (41.1) can be written

$$[(h+\sqrt{h^2-ab})x+by+g+e\sqrt{g^2-b\bar{c}}\,][(h-\sqrt{h^2-ab})x+by+g - e\sqrt{g^2-b\bar{c}}\,]=b(\bar{c}-c),$$

where e is $+1$ or -1 according as $hg-bf>0$ or <0. Show that for the hyperbola (40.14) $c=-3$, and obtain by this method the results following (40.15).

11. Show that $x^2-2xy+2y^2-2x+4y=0$ is an equation of an ellipse, and obtain its equation in the form (40.10) by the method of Ex. 10.

42. Center, Principal Axes, and Tangents of a Conic Defined by a General Equation

In accordance with Theorem [**6.3**],

$$x=x_1+tu, \qquad y=y_1+tv \tag{42.1}$$

are parametric equations of the line through the point (x_1, y_1) and with direction numbers u and v, t being proportional to the directed distance from (x_1, y_1) to a representative point (x, y). In order to find the values of t, if any, for points in which the line so defined meets the curve

$$ax^2+2hxy+by^2+2fx+2gy+c=0, \tag{42.2}$$

we substitute the above expressions for x and y and obtain the equation

(42.3)

$$(au^2+2huv+bv^2)t^2+2[(ax_1+hy_1+f)u+(hx_1+by_1+g)v]t + (ax_1^2+2hx_1y_1+by_1^2+2fx_1+2gy_1+c)=0.$$

We make use of this result in the discussion of a number of questions.

If (x_1, y_1) is to be the mid-point of the chord of the conic with direction numbers u and v, that is, the line segment with the points of intersection of the line and the conic as end points, the two solutions of (42.3) must differ only in sign, and hence the coefficient of t in (42.3) must be equal to zero; that is,

$$(ax_1+hy_1+f)u+(hx_1+by_1+g)v=0. \tag{42.4}$$

In order that (x_1, y_1) shall be the center of the conic, it must be the mid-point of every chord through it, and consequently (42.4) must hold for every value of u and v. This means that the center (x_0, y_0), if the conic has a center, is given by

$$ax_0 + hy_0 + f = 0, \qquad hx_0 + by_0 + g = 0. \tag{42.5}$$

By Theorem [9.1] these equations have one and only one common solution if $ab - h^2 \neq 0$. When this condition is satisfied, the coordinates of the center are given by (see § 41, Ex. 6)

$$x_0 = \frac{hg - bf}{ab - h^2}, \qquad y_0 = \frac{hf - ag}{ab - h^2}. \tag{42.6}$$

If the conic (42.2) is not degenerate, these are the coordinates of the center of an ellipse when $ab - h^2 > 0$, and of the center of a hyperbola when $ab - h^2 < 0$.

If the conic is degenerate and $ab - h^2 \neq 0$, and (x_0, y_0) is now the intersection of the two real or imaginary lines, then the expression in the last parentheses in (42.3) is equal to zero. Also, if u and v are direction numbers of either of the lines, equation (42.3) must be satisfied for every value of t. Consequently we have (42.4) holding, and also

$$au^2 + 2\,huv + bv^2 = 0.$$

From this it follows, as shown before, that the lines are real or imaginary according as $ab - h^2 < 0$ or > 0. Since (42.4) must hold for the direction numbers of the two lines, we again obtain equations (42.5). Consequently the coordinates of the vertex, that is, the point of intersection of the lines, are given by (42.6). When equation (42.2) is written in the form

$$(ax + hy + f)x + (hx + by + g)y + fx + gy + c = 0, \tag{42.7}$$

it follows from (42.5) that

$$fx_0 + gy_0 + c = 0. \tag{42.8}$$

This equation and equations (42.5) have a common solution, if and only if $D = 0$, where D is defined by (41.10); this result is in accord with Theorem [**41.4**]. Hence we have

[42.1] *When $ab - h^2 \neq 0$, the conic* (42.2) *is a central conic with center* (42.6) *when $D \neq 0$, and is a degenerate conic with vertex* (42.6) *when $D = 0$.*

When $ab - h^2 = 0$, by Theorem [**9.3**] equations (42.5) do not have a common solution unless $hg - bf = 0$ and $hf - ag = 0$, in which case the conic is degenerate, as follows from (41.9) and (41.11). Thus, as stated in § 35, a parabola does not have a center. When the conic is degenerate, and consists of two parallel or coincident lines, equations (42.5) are equivalent; they are equations of a line of centers, that is, a line of points of symmetry.

Returning to the consideration of equation (42.4), we observe that for fixed values of u and v, any point of the line

$$(42.9) \qquad (ax + hy + f)u + (hx + by + g)v = 0,$$

for which the values of t given by (42.3) are real when x_1 and y_1 are the coordinates of such a point, is the mid-point of the chord through this point and with direction numbers u and v. Hence we have

[**42.2**] *Equation* (42.9) *is an equation of the locus of the mid-points of the set of parallel chords with direction numbers* u *and* v.

In order that (42.9) shall be an equation of a principal axis, that is, a line of symmetry of the conic (42.2), the line (42.9) must be perpendicular to the chords it bisects; consequently u and v must be such that they are direction numbers of the perpendiculars to the line. When equation (42.9) is written

$$(au + hv)x + (hu + bv)y + (fu + gv) = 0,$$

we observe that $au + hv$ and $hu + bv$ are direction numbers of any line perpendicular to (42.9) by Theorem [**6.9**]. Consequently u and v must be such that

$$au + hv = ru, \qquad hu + bv = rv,$$

where r is a factor of proportionality; $r \neq 0$, otherwise there is no line (42.9). When these equations are written

$$(42.10) \qquad (a - r)u + hv = 0, \qquad hu + (b - r)v = 0,$$

we have that a solution u, v, not both zero, of these equations is given for each value of r, other than zero, satisfying the equation

$$(42.11) \qquad \begin{vmatrix} a - r & h \\ h & b - r \end{vmatrix} = 0,$$

as the reader can readily show. When this equation is expanded, it becomes

$$(42.12) \qquad r^2 - (a + b)r + (ab - h^2) = 0.$$

This equation is called the *characteristic equation* of equation (42.2). We consider the two cases when $ab - h^2 = 0$ and $ab - h^2 \neq 0$.

Case 1. $\mathbf{ab - h^2 = 0.}$ In this case $r = a + b$ is the only non-zero root, and from (42.10) we have

$$\frac{u}{v} = \frac{h}{b} = \frac{a}{h}.$$

In consequence of this result equation (42.9) reduces to

$$(42.13) \qquad (a + b)(hx + by) + fh + gb = 0.$$

Hence we have (see § 41, Ex. 5)

[42.3] *When* (42.2) *is an equation of a parabola,* (42.13) *is an equation of the principal axis of the parabola.*

Case 2. $\mathbf{ab - h^2 \neq 0.}$ In this case the solutions of (42.12) are found by the use of the quadratic formula to be

$$(42.14) \qquad r = \frac{(a + b) \pm \sqrt{(a - b)^2 + 4h^2}}{2},$$

and are always real numbers. The two roots are different, unless $a = b$ and $h = 0$, in which case the conic is a circle (see § 12). In any particular case, other than that of a circle, with the two values of r from (42.14) two sets of values of u and v may be found from (42.10), which when substituted in (42.9) give the two principal axes of the conic.

When the left-hand member of (42.2) is the product of two factors of the first degree, in which case by Theorem [**41.5**] $D = 0$, and a transformation of coordinates is effected, the resulting expression in x', y' is the product of two factors of the first degree, and consequently $D' = 0$ for this expression, where D' is the corresponding function (41.10). From this result it does not follow necessarily that $D = D'$ in general, but we shall show that this is true, and consequently that D is an invariant under any transformation of rectangular coordinates. We prove this not by direct substitution, but by the following interesting device. We denote by $f(x, y)$ the left-hand member of (42.2), and consider the expression

$$f(x, y) - r(x^2 + y^2 + 1), \tag{42.15}$$

that is,

$$(a - r)x^2 + 2\,hxy + (b - r)y^2 + 2\,fx + 2\,gy + (c - r).$$

By Theorem [**41.5**] the condition that this expression shall be the product of two factors of the first degree is that r shall be such that

$$\begin{vmatrix} a - r & h & f \\ h & b - r & g \\ f & g & c - r \end{vmatrix} = 0.$$

This equation, upon expansion of the determinant, is

$$r^3 - Ir^2 + Jr - D = 0, \tag{42.16}$$

where

$$I = a + b + c, \qquad J = ab + bc + ac - h^2 - f^2 - g^2. \tag{42.17}$$

When a rotation of the axes (28.8) is applied to (42.15), we obtain

$$f'(x', y') - r(x'^2 + y'^2 + 1), \tag{42.18}$$

where $f'(x', y')$ denotes the *transform* of $f(x, y)$, that is, the expression into which $f(x, y)$ is transformed. If (42.15) is the product of two factors of the first degree, so also is (42.18) and we have

$$r^3 - I'r^2 + J'r - D' = 0,$$

where I', J', and D' are the same functions of $a', \cdots, c'$ as I, J, and D respectively are of $a, \cdots, c$. Since r is not affected by the transformation, it follows that $I = I'$, $J = J'$, $D = D'$. Consequently I, J, and D are invariants under any rotation of the axes.

When a translation (28.1) is applied to $f(x, y)$, a, b, and h are not changed, and we have

$$\begin{aligned} f'(x', y') = ax'^2 + by'^2 + 2\,hx'y' &+ 2(ax_0 + hy_0 + f)x' \\ &+ 2(hx_0 + by_0 + g)y' + f(x_0, y_0). \end{aligned}$$

For this expression D' is

$$\begin{vmatrix} a & h & ax_0 + hy_0 + f \\ h & b & hx_0 + by_0 + g \\ ax_0 + hy_0 + f & hx_0 + by_0 + g & f(x_0, y_0) \end{vmatrix}.$$

If, considering $f(x_0, y_0)$ expressed in the form (42.7), we subtract from the last row the first row multiplied by x_0 and the second multiplied by y_0, we have

$$\begin{vmatrix} a & h & ax_0 + hy_0 + f \\ h & b & hx_0 + by_0 + g \\ f & g & fx_0 + gy_0 + c \end{vmatrix},$$

which is seen to be equal to D. Since any transformation of rectangular coordinates is equivalent to a rotation and a translation, we have

[42.4] *The function D of the coefficients of an equation of the second degree in x and y is an invariant under any change of rectangular coordinate axes.*

Since, as shown in § 40, $a + b$ and $ab - h^2$ are invariants, it follows that the roots of the characteristic equation are the same in every coordinate system. When by a rotation of the axes equation (42.2) is transformed into an equation without a term in xy, it follows from (42.11) that the coefficients of the terms in x^2 and y^2 are the roots of the characteristic equation. Hence, when one finds the roots of the characteristic equation of an equation (42.2), one has obtained the numbers which are the coefficients of the second-degree terms in an equation without a term in xy into which (42.2) is transformable.

In accordance with the theory of algebraic equations it follows from (42.12) that

$$ab - h^2 = r_1 r_2, \tag{42.19}$$

where r_1, r_2 are the roots of (42.11). We consider the cases when $ab - h^2 \neq 0$ and $ab - h^2 = 0$.

Case 1. $ab - h^2 \neq 0$. From (42.19) it follows that both roots are different from zero, and from the results of § 40 that (42.2) is transformable into an equation of the form

$$ax^2 + by^2 + c = 0.$$

In this case $r_1 = a$, $r_2 = b$, and $D = abc = r_1 r_2 c$. Consequently in an appropriate coordinate system an equation of the conic is

$$r_1 x^2 + r_2 y^2 + \frac{D}{r_1 r_2} = 0. \tag{42.20}$$

Hence we have

[42.5] *Equation* (42.2) *for which* $ab - h^2 \neq 0$ *is an equation of an ellipse or of a hyperbola, when* $D \neq 0$, *according as the roots have the same or opposite signs; it is degenerate when* $D = 0$.

Case 2. $ab - h^2 = 0$. From (42.19) it follows that at least one of the roots is zero, and the reader can show that the other is not zero (see Ex. 6). From the results of § 40 it follows that (42.2) is transformable into an equation of the form

$$ax^2 + 2\,by = 0 \qquad \text{or} \qquad ax^2 + c = 0.$$

In both cases we may take $r_1 = a$, $r_2 = 0$. In the first case $D = -r_1 b^2$; in the second case $D = 0$. Hence an equation of the conic is

$$r_1 x^2 + 2\sqrt{-\frac{D}{r_1}}\,y = 0 \qquad \text{or} \qquad r_1 x^2 + c = 0. \tag{42.21}$$

The second equation is an equation of two parallel or coincident lines according as $c \neq 0$ or $c = 0$. In the latter case every point of the coincident lines is a point of symmetry and lies on the locus. Consequently, if c in the second of equations (42.21) is to be zero, the common solutions of equations (42.5) for a given equation (42.2) must be solutions of the given equation.

The preceding results are set forth in the following table, in which the *canonical*, or type, form of an equation of a conic is given in terms of the roots of its characteristic equation:

$ab - h^2 \neq 0$	$D \neq 0$	$r_1 x^2 + r_2 y^2 + \frac{D}{r_1 r_2} = 0$ . . .	Ellipse or hyperbola
	$D = 0$	$r_1 x^2 + r_2 y^2 = 0$	Two lines
$ab - h^2 = 0$	$D \neq 0$	$r_1 x^2 + 2\sqrt{-\frac{D}{r_1}}\,y = 0$ . . .	Parabola
	$D = 0$	$r_1 x^2 + c = 0$*	Two parallel or coincident lines

*$c = 0$ if (42.5) and (42.2) have a common solution.

When in a particular problem one has found the values of the roots, he is able to determine completely the form of the conic, but in order to find its position relative to the given coordinate axes it is necessary that he find the principal axes, or axis, by means of (42.9) and (42.10).

We consider finally the case when (x_1, y_1) is a point of the conic and (42.1) are equations of the tangent to the conic at this point. Since the point (x_1, y_1) is on the conic, the expression in the last parentheses in (42.3) is equal to zero. Consequently one solution of (42.3) is $t = 0$, for which from (42.1) we have the point (x_1, y_1). The other solution of (42.3) when substituted in (42.1) gives the coordinates of the other point in which the line meets the conic. If the line is to be tangent at (x_1, y_1), this other solution also must be zero, that is, u and v must be such that

$$(ax_1 + hy_1 + f)u + (hx_1 + by_1 + g)v = 0.$$

From this it follows that $hx_1 + by_1 + g$ and $-(ax_1 + hy_1 + f)$, being proportional to u and v, are direction numbers of the tangent, and consequently an equation of the tangent at the point (x_1, y_1) is

$$\frac{x - x_1}{hx_1 + by_1 + g} + \frac{y - y_1}{ax_1 + hy_1 + f} = 0.$$

When this equation is cleared of fractions, and the expression $ax_1^2 + 2hx_1y_1 + by_1^2$ is replaced by $-(2fx_1 + 2gy_1 + c)$, to which it is equal, as follows from (42.2), we have

[42.6] *An equation of the tangent to the conic* (42.2) *at the point* (x_1, y_1) *is*

$$ax_1x + h(y_1x + x_1y) + by_1y + f(x + x_1) + g(y + y_1) + c = 0. \tag{42.22}$$

EXERCISES

1. Find the axis, vertex, and tangent at the vertex of the parabola

$$4x^2 - 12xy + 9y^2 - 3x - 2y + 4 = 0.$$

2. Find the principal axes and center of the following:

a. $8x^2 - 4xy + 5y^2 - 36x + 18y + 9 = 0.$

b. $2x^2 - 4xy - y^2 + 7x - 2y + 3 = 0.$

Show that the axes are perpendicular to one another.

3. Show that when (42.2) is an equation of a central conic, and a translation of axes is effected by the equations $x = x' + x_0$, $y = y' + y_0$, for which the point (x_0, y_0) is the center, in the resulting equation there are no terms of the first degree in x' and y'. Apply this process to the equations in Ex. 2.

4. Determine whether equations (34.5), (36.7), and (36.7′) conform to Theorem [**42.6**], and formulate a rule for obtaining (42.22) from (42.2).

5. Find equations of the tangent and normal at the point (2, − 1) to the conic
$$2x^2 - 4xy + 3y^2 - 2x + 3y - 12 = 0.$$

6. Show that not both of the roots of the characteristic equation (42.12) can be zero.

7. By finding the roots of the characteristic equation determine completely the form of each of the conics in Ex. 2 and also in § 40, Ex. 1.

8. What is the character of the locus of equation (42.2), for which $ab - h^2 = 0$, when it is possible to derive by a translation of axes an equation in which there are no terms of the first degree in x' and y'?

9. Prove that the centers of the conics whose equations are
$$ax^2 + 2hxy + by^2 + 2ftx + 2gty + c = 0,$$
as the parameter t takes different values, lie on a straight line through the origin.

43. Locus Problems

In the closing paragraphs of § 13 we explained what is meant by finding the locus of a point satisfying certain geometric conditions. In § 32 it was stated that when a locus is defined geometrically and without reference to a coordinate system, the reader is free to choose any set of coordinate axes in obtaining an equation of the locus, but that it is advisable that axes be chosen in such manner relative to the geometric configuration as to obtain an equation in simple form. In some cases the choice of an axis of symmetry, if there is such, as one of the coordinate axes tends toward simplicity in the equation. We give below two examples to be studied before the reader proceeds to the solution of the exercises at the end of this section. Before doing so he should understand that in certain cases only a portion of the graph of an equation obtained for a given problem is the locus, that is, some points of the graph do not satisfy the geometric definition of the locus; and the reader must have this in mind as he interprets geometrically any equation he has obtained.

1. *Find the locus of the vertex of a triangle whose base is fixed in position and length, and whose angles are such that the product of the tangents of the base angles is a constant, not zero.*

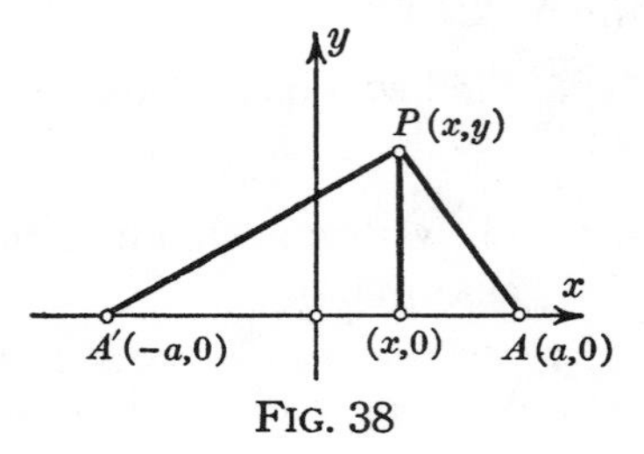

FIG. 38

We choose the base of the triangle for the x-axis and its mid-point for origin. If we denote by $2a$ the length of the base, its end points are $A'(-a, 0)$ and $A(a, 0)$. In accordance with the problem (see Fig. 38),

$$\tan PA'A \cdot \tan PAA' = k \ (\neq 0).$$

If, as in the figure, the base angles are acute, the constant k is positive. If P lies so that one of the base angles is obtuse, k is negative. In both cases we have from the above equation

$$\frac{y}{x+a} \cdot \frac{y}{a-x} = k.$$

When P lies below the x-axis, each y in the equation must be replaced by $-y$, but this does not change the equation. All such considerations as these must be taken into account in studying a problem, if one is to be sure that one is handling every aspect of it.

When the above equation is written in the form

$$\frac{x^2}{a^2} + \frac{y^2}{ka^2} = 1,$$

it is seen that when $k > 0$ the curve is an ellipse, and that $A'A$ is the major or minor axis according as $k < 1$ or $k > 1$; and that when $k < 0$, the curve is a hyperbola with $A'A$ for transverse axis. In both cases all points of the curve except A' and A satisfy the definition of the locus.

2. *The base of a triangle is given in length and position, and one of the angles at the base is double the other; find the locus of the vertex.*

We take the base as the x-axis and its mid-point as origin, with the result that the coordinates of the ends of the base are of the form $(-a, 0)$ and $(a, 0)$. We denote the angle at $(-a, 0)$ by ϕ and the angle at $(a, 0)$ by 2ϕ (see Fig. 39). In accordance with the definition of the locus we have

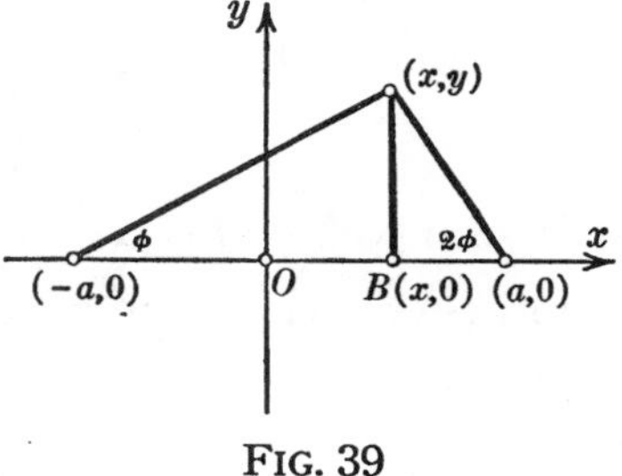

FIG. 39

$$(43.1) \qquad \tan\phi = \frac{y}{a+x}, \qquad \tan 2\phi = \frac{y}{a-x}.$$

Dividing the first equation by the second and solving for x, we obtain

$$(43.2) \qquad x = \frac{a(\tan 2\phi - \tan\phi)}{\tan 2\phi + \tan\phi}.$$

From this result and the first of (43.1) we obtain

$$(43.3) \qquad y = \frac{2a\tan\phi\tan 2\phi}{\tan 2\phi + \tan\phi}.$$

Thus x and y are expressed in terms of ϕ as a *parameter*, and equations (43.2) and (43.3) are *parametric equations* of the locus.

In many locus problems, particularly those in which the point P is defined with respect to movable points and lines, it is advisable to use a parameter, and in some cases several parameters which are not independent but connected by $n-1$ relations, n being the number of parameters. By eliminating the parameters from these equations and from the two expressions for x and y in terms of these, we obtain an equation in x and y of the locus.

In the above problem the equation of the curve in x and y is obtained by eliminating ϕ from (43.2) and (43.3), or more readily from (43.1) by using the formula from trigonometry $\tan 2\phi = 2\tan\phi/(1-\tan^2\phi)$. This gives the equation

$$y(3x^2 - y^2 + 2ax - a^2) = 0;$$

this equation is equivalent to $y = 0$, the x-axis, which is an evident solution of the problem, and to the equation obtained by equating to zero the expression in parentheses. When this equation is written

$$3\left(x + \frac{a}{3}\right)^2 - y^2 = \frac{4\,a^2}{3}, \tag{43.4}$$

it is seen to be an equation of a hyperbola with center at the point $\left(-\frac{a}{3}, 0\right)$, vertices at $(-a, 0)$ and $\left(\frac{a}{3}, 0\right)$, and semi-conjugate axis $2\,a/\sqrt{3}$. Only the branch of the hyperbola through the vertex $\left(\frac{a}{3}, 0\right)$ satisfies the conditions of the problem, as one sees geometrically.

This problem is of historical interest because of its relation to that of the trisection of an angle, a problem which goes back to the Greeks, who endeavored to obtain a construction for trisecting an angle, using only a ruler and a compass. Years ago it was shown by algebraic considerations that this is impossible for a general angle, but many people keep on trying to do it. We shall show that it can be done by means of a hyperbola constructed not point by point but continuously (see § 35 after Theorem [**35.1′**]).

Suppose that on a line $A'(-a, 0)A(a, 0)$ the hyperbola (43.4) has been accurately constructed, as shown in Fig. 40. Through the point A' and below the line $A'A$ a line is drawn making an angle of $90° - \theta$ with the line $A'A$, θ being a given acute angle. Denote by C the point of intersection of this line and the perpendicular to $A'A$ at its midpoint O. With C as center and CA' as radius describe a circle, meeting the branch of the hyperbola through $\left(\frac{a}{3}, 0\right)$ in the point P. The angle $A'CA$ is equal to $2\,\theta$; consequently the angle $A'PA$ is $\frac{1}{2}(360° - 2\,\theta) = 180° - \theta$. But by the above problem the angle $A'PA$ is $180° - 3\,\phi$. Consequently $\phi = \frac{1}{3}\,\theta$, as was to be shown.

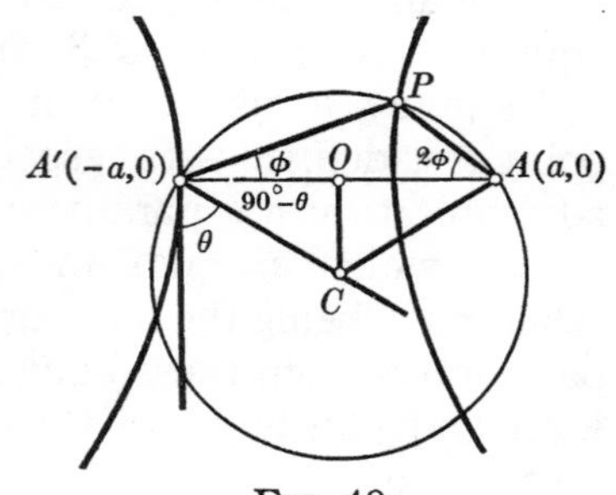

FIG. 40

From this result it follows that if a hyperbola could be constructed continuously, instead of only point by point, by means of a ruler and compass, a given angle could be trisected by this means; since it has been shown, as remarked above, that the latter is impossible, it follows that a hyperbola cannot be constructed continuously by means of ruler and compass, although it can be by other means (see § 35).

EXERCISES

1. Find the locus of a point the sum of the squares of whose distances from two fixed points is constant.

2. Find the locus of a point whose distances from two fixed points are in constant ratio.

3. Find the locus of a point the square of whose distance from a fixed point is a constant times its distance from a fixed line not passing through the fixed point.

4. Find the locus of a point the sum of the squares of whose distances from two intersecting lines is constant.

5. Find the locus of a point such that the square of its distance from the base of an isosceles triangle is equal to the product of its distances from the other two sides.

6. Find the locus of a point such that the length of a tangent drawn from it to one of two given circles is a constant times the length of a tangent drawn from it to the other circle.

7. Find the locus of a point the sum of whose distances from two fixed perpendicular lines is equal to the square of its distance from the point of intersection of the lines.

8. Find the locus of a point the sum of the squares of whose distances from two fixed perpendicular lines is equal to the square of its distance from the point of intersection of the lines.

9. Find the locus of a point the product of whose distances from two fixed intersecting lines is constant.

10. Find the locus of a point the sum of the squares of whose distances from two adjacent sides of a square is equal to the sum of the squares of its distances from the other two sides.

11. Find the locus of a point which is the center of a circle passing through a fixed point and tangent to a fixed line.

12. Find the locus of a point which is the center of a circle tangent to a fixed line and to a fixed circle.

13. Given two parallel lines L_1 and L_2, and a third line L_3 perpendicular to the first two, find the locus of a point the product of whose distances from L_1 and L_2 is a constant times the square of its distance from L_3.

14. Given a circle which is tangent to a given line L at the point A; denote by B the other end of the diameter through A; through A draw a line and denote by Q and R its points of intersection with the circle and with the tangent to the circle at B. The locus of the point of intersection of a line through Q parallel to the line L and of a line through R parallel to the diameter AB as the line through A is rotated about A is called the *witch of Agnesi.* Find an equation of the locus and draw its graph.

15. Through the point (2, 0) a line is drawn meeting the lines $y = x$ and $y = 3x$ in the points A and B. Find the locus of the mid-point of AB for all the lines through (2, 0).

16. A variable line makes with two fixed perpendicular lines a triangle of constant area. Find the locus of the point dividing in constant ratio the segment of the variable line whose end points are on the two fixed lines.

17. A variable line is drawn parallel to the base BC of a triangle ABC, meeting AB and AC in the points D and E respectively. Find the locus of the intersection of BE and CD.

18. A set of parallel line segments are drawn with their ends on two fixed perpendicular lines. Find the locus of the point dividing them in the ratio $h : k$.

19. One side of each of a set of triangles is fixed in position and length, and the opposite angle is of fixed size. Find the locus of the centers of the inscribed circles of these triangles.

20. Find the locus of the intersection of the diagonals of rectangles inscribed in a given triangle, one side of each rectangle being on the same side of the triangle.

21. In a rectangle $ABCD$ line segments EF and GH are drawn parallel to AB and BC respectively and with their end points on the sides of the rectangle. Find the locus of the intersection of HF and EG.

22. Find the locus of the mid-points of the chords of a circle which meet in a point on the circle.

23. Given the base AB of a triangle ABC, find the locus of the vertex C,

(*a*) when $\overline{CA}^2 - \overline{CB}^2$ is constant;
(*b*) when $\overline{CA}^2 + \overline{CB}^2$ is constant;
(*c*) when CA/CB is constant;
(*d*) when the angle at C is constant;
(*e*) when the difference of the base angles A and B is constant.

24. Given the base AB and the opposite angle C of the triangle ABC, find the locus of the point of intersection of the perpendiculars from A and B upon the opposite sides.

25. AB is a fixed chord of a circle and C is any point of the circle. Find the locus of the intersection of the medians of the triangles ABC.

26. AB is a line segment of fixed length and position, and C is any point on a line parallel to AB. Find the locus of the intersection of the three altitudes of the triangle ABC; also the locus of the intersection of the medians.

27. Through each of two fixed points $P_1(x_1, y_1)$ and $P_2(x_2, y_2)$ lines are drawn perpendicular to one another; denote by A and B the points in which these lines meet the y-axis and x-axis respectively. Find the locus of the mid-point of AB. (Use the slope of either line as parameter.)

28. Show that the locus of a point the tangents from which to an ellipse are perpendicular to one another is a circle. This circle is called the *director circle* of the ellipse.

29. Find the locus of a point the tangents from which to the parabola $y^2 = 4\,ax$ include an angle of 45°. (Use the slopes m_1 and m_2 of the tangents as parameters.)

30. Find the locus of the extremities of the minor axes of the ellipses which have a given point for focus and a given line for directrix.

31. Find an equation of the locus of a point the product of whose distances from two fixed points is a constant k^2. Observe that as k^2 takes on different values, the curve varies in form; these curves are known as the *ovals of Cassini*. When k is equal to half the distance between the points, the curve is the lemniscate (see § 29, Ex. 1).

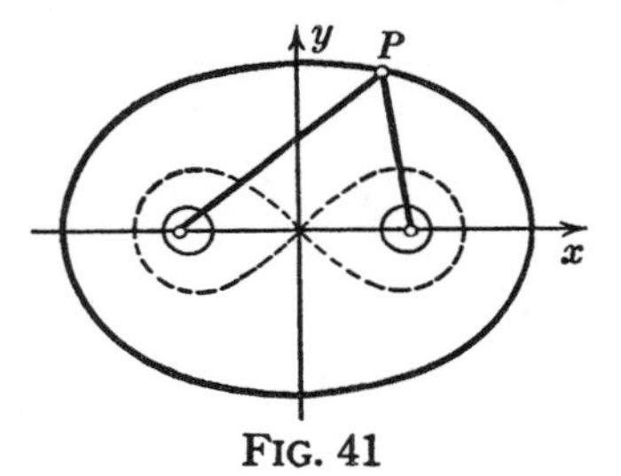

FIG. 41

32. Find the parametric equations of the *cycloid,* the locus described by a point P on the circumference of a circle as the circle rolls along a fixed line, using as parameter the angle ϕ at the center C of the circle formed by the line CP and the perpendicular to the line on which the circle rolls.

33. Through each of two fixed points P_1 and P_2 lines are drawn so as to intercept a constant length on a fixed line below P_1 and P_2; find the locus of the intersection of the variable lines. Can the line P_1P_2 be parallel to the fixed line?

34. Find the locus of the points of contact of the tangents drawn to a set of confocal ellipses from a fixed point on the line of the major axes of the ellipses (see § 36, Ex. 14).

35. Given two concentric ellipses one within the other, and with their principal axes on the same lines, if P is the pole with respect to the outer ellipse of a line tangent to the inner ellipse, find the locus of P.

36. Find the locus of the centers of the conics

$$ax^2 + 2\,hxy + by^2 + 2\,gy = 0,$$

when a, h, and g are fixed and b varies (see (42.6)).

"

CHAPTER 5

The Quadric Surfaces

"

44. Surfaces of Revolution. The Quadric Surfaces of Revolution

When one thinks of a surface, among the first examples which come to mind are the plane and the sphere. In § 17 we showed that any equation of the first degree in x, y, z is an equation of a plane. In consequence of (15.2) an equation of a sphere of radius a and center at the origin is $x^2+y^2+z^2=a^2$. In these two cases the coordinates of any point of the surface are a solution of a single equation in the coordinates. Whether we use rectangular coordinates, or spherical coordinates, or any other, we say that the locus of a point in space whose coordinates satisfy a single equation is a *surface*. When rectangular coordinates are used and one of the coordinates does not appear in the equation, the locus is a cylinder whose elements are parallel to the axis of this coordinate. For example, if z does not appear in the equation, the locus is a cylinder whose elements are parallel to the z-axis. In fact, if (x_1, y_1, z_1) is any point on the surface, so also is (x_1, y_1, z_2), where z_2 is any z, because the equation imposes a condition upon x and y and none upon z, and after x and y have been chosen to satisfy the equation these two and any z are coordinates satisfying the equation.

A line in space may be thought of as the intersection of two planes. In fact, in rectangular coordinates a line is defined by two equations of the first degree in x, y, and z, as was done in § 19. Thus a line is a special case of a *curve*, which may be defined as the intersection of two surfaces; that is, a curve in space is defined by two independent equations. In the xy-plane the circle with center at the origin and radius a has as an equation $x^2+y^2=a^2$, but when thought of as a curve in space it has in addition the equation $z=0$. From the viewpoint of space the former of these equations is a circular cylinder with the z-axis for its axis, and consequently the circle under consideration is the intersection of this cylinder and the plane $z=0$.

Ordinarily the curve of intersection of two surfaces, neither of which is a plane, does not lie entirely in a plane. Such a curve is called a *twisted*, or *skew*, curve, and one which lies entirely in a plane is called a *plane* curve. Consider, for example, the curve defined in terms of a parameter t by the equations

$$x = t, \quad y = t^2, \quad z = t^3. \tag{44.1}$$

This curve meets the plane

$$ax + by + cz + d = 0$$

in the points for which the parameter t has values which satisfy the equation

$$at + bt^2 + ct^3 + d = 0.$$

Consequently at most three of the points of the curve lie in any one plane. The curve defined by (44.1) is called a *twisted cubic.*

When a plane curve in space is revolved about a line in its plane, the surface generated is called a *surface of revolution.* For example, when a circle is revolved about a diameter, the surface generated is a sphere. If a curve lying in the xy-plane is revolved about the x-axis, each point of the curve describes a circle with center on the x-axis. Let $P(x, y, z)$ be a representative point of such a surface; for all other points of the same circle as P the coordinate x is the same and y and z are different, but they are such that $\sqrt{y^2 + z^2}$ is the same, since it is the radius of the circle. When P is in the xy-plane, y is the radius and $z = 0$. Hence an equation of the surface is obtained from an equation of the curve in the plane $z = 0$ on replacing y by $\sqrt{y^2 + z^2}$.

We apply this process to the parabola with the equation $y^2 = 4\,ax$ in the plane $z = 0$, and get the equation

$$y^2 + z^2 = 4\,ax; \tag{44.2}$$

the surface generated is called a *paraboloid of revolution.* If the parabola is revolved about the y-axis, by an argument similar to the above we have

$$y^2 = 4\,a\sqrt{x^2 + z^2}.$$

Squaring both sides, we obtain the equation of the fourth degree

$$y^4 = 16\,a^2(x^2 + z^2),$$

and consequently the surface generated is said to be a surface of the fourth degree. The preceding examples illustrate the fact that when a curve is symmetric with respect to the line about which it is revolved, the degree of the surface is the same as the degree of the curve, but when the curve is not symmetric with respect to this line, the degree of the surface is twice that of the curve. Another example of the latter case is afforded by the rotation of the line $ax + by = 0$ about the x-axis. The surface generated is a cone, consisting of what the reader might consider as two right circular cones with the x-axis as common axis, an element of one cone being the prolongation through its vertex of an element of the other. An equation of the surface is

$$a^2x^2 = b^2(y^2 + z^2). \tag{44.3}$$

According as the ellipse $\frac{x^2}{a^2} + \frac{y^2}{b^2} = 1$, $z = 0$ is revolved about the x-axis or the y-axis, the surface generated has as an equation

$$\frac{x^2}{a^2} + \frac{y^2 + z^2}{b^2} = 1 \quad \text{or} \quad \frac{x^2 + z^2}{a^2} + \frac{y^2}{b^2} = 1. \tag{44.4}$$

When $a > b$, the first surface has a shape somewhat like a football, and is called a *prolate spheroid*; the second surface is discus-like in shape, particularly when b is much smaller than a, and is called an *oblate spheroid*; when $a < b$, the situation is reversed.

According as the hyperbola $\frac{x^2}{a^2} - \frac{y^2}{b^2} = 1$, $z = 0$ is revolved about the x-axis or the y-axis, the surface generated has as an equation

$$\frac{x^2}{a^2} - \frac{y^2 + z^2}{b^2} = 1 \quad \text{or} \quad \frac{x^2 + z^2}{a^2} - \frac{y^2}{b^2} = 1. \tag{44.5}$$

The first surface consists of two parts analogous to Fig. 45, and is called a *hyperboloid of revolution of two sheets*; the second surface is shaped somewhat like a spool of endless extent, and is called a *hyperboloid of revolution of one sheet*. When the asymptotes are rotated simultaneously with the hyperbola, they generate cones with the respective equations

$$\frac{x^2}{a^2} - \frac{y^2 + z^2}{b^2} = 0, \quad \frac{x^2 + z^2}{a^2} - \frac{y^2}{b^2} = 0. \tag{44.6}$$

In the first case the two sheets of the hyperboloid lie inside the two parts of the cone, one in each part; in the second case the cone lies inside the hyperboloid.

EXERCISES

1. Show that the curve (44.1) is the intersection of the cylinders $y = x^2$, $z = x^3$, $y^3 = z^2$.

2. Show that for each value of t the corresponding plane

$$3\,t^2x - 3\,ty + z - t^3 = 0$$

meets the curve (44.1) in three coincident points, and that ordinarily through a point in space not on this curve there are three planes each of which meets the curve in three coincident points.

3. Show that the curve $x = a \cos t$, $y = a \sin t$, $z = bt$ lies on the cylinder $x^2 + y^2 = a^2$, and that it is not a plane curve.

4. Find an equation of the cone generated by revolving the line $2\,x - 3\,y + 6 = 0$ about the x-axis. Find the vertex of the cone and the curve in which the cone intersects the yz-plane.

5. Find an equation of the *torus* generated by revolving about the y-axis the curve $(x - a)^2 + y^2 - b^2 = 0$, $z = 0$, where $b < a$.

6. Find the locus of a point equidistant from the point $(a, 0, 0)$ and the plane $x + a = 0$.

7. Find the locus of a point equidistant from the z-axis and the plane $z = 0$.

8. Find the locus of a point the sum of whose distances from the points $(c, 0, 0)$ and $(-c, 0, 0)$ is a constant; also the locus when the difference of these distances is a constant.

9. Find the points in which the surface $x^2 + y^2 - z^2 + 7 = 0$ is met by the line

$$\text{(i)} \qquad \frac{x-1}{2} = \frac{y-2}{3} = \frac{z+1}{2} = t,$$

by finding the values of t for which x, y, and z given by equations (i) are coordinates of a point on the surface.

10. Find the points in which the surface $x^2 - 2\,xy + 3\,z^2 - 5\,y + 10 = 0$ is met by the line

$$\frac{x-3}{1} = \frac{y+2}{-2} = \frac{z-3}{2},$$

and interpret the result.

11. Show that the curve of intersection C of the quadric

$$x^2 + y^2 - 5\,z^2 - 5 = 0$$

and the plane $x = 3\,z$ lies on the cylinders $y^2 + 4\,z^2 - 5 = 0$ and $4\,x^2 + 9\,y^2 - 45 = 0$. What are the respective projections of C on the coordinate planes, and what kind of curve is C?

12. Show that the line $x - 1 = y - 2 = z + 1$ lies entirely on the surface $z^2 - xy + 2x + y + 2z - 1 = 0$.

13. Find equations of the projections upon each of the coordinate planes of the curve of intersection of the plane $x - y + 2z - 4 = 0$ and the surface $x^2 - yz + 3x = 0$.

45. Canonical Equations of the Quadric Surfaces

On interchanging x and z in (44.2), the resulting equation is a special case of

$$ax^2 + by^2 = cz, \tag{45.1}$$

in which a and b have the same sign. A plane $z = k$, where k is a constant, intersects the surface in the curve whose equations are

$$ax^2 + by^2 = ck, \quad z = k. \tag{45.2}$$

The first of these equations is an equation of an elliptical cylinder, and consequently the curve is an ellipse. In like manner a plane $y = k$ intersects the surface in the curve

$$ax^2 = c\left(z - \frac{bk^2}{c}\right), \quad y = k, \tag{45.3}$$

which is a parabola; a similar result follows for a plane $x = k$. When a and b in (45.1) differ in sign, the curve of intersection (45.2) is a hyperbola, and the curve (45.3) is a parabola. The surface (45.1) is called an *elliptic paraboloid* or a *hyperbolic paraboloid* according as a and b have the same or different signs; these surfaces are illustrated in Figs. 42 and 43 respectively.

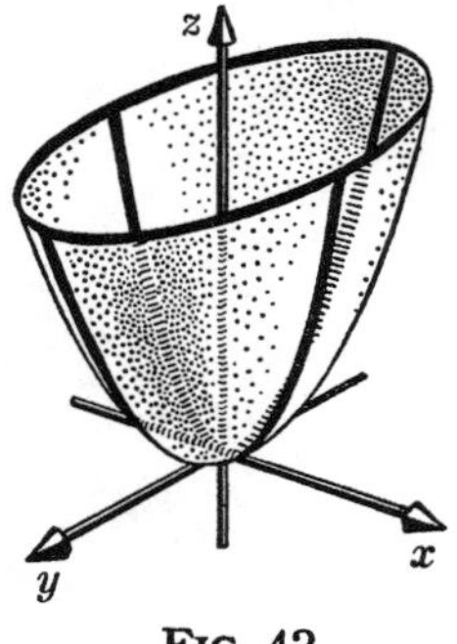

FIG. 42

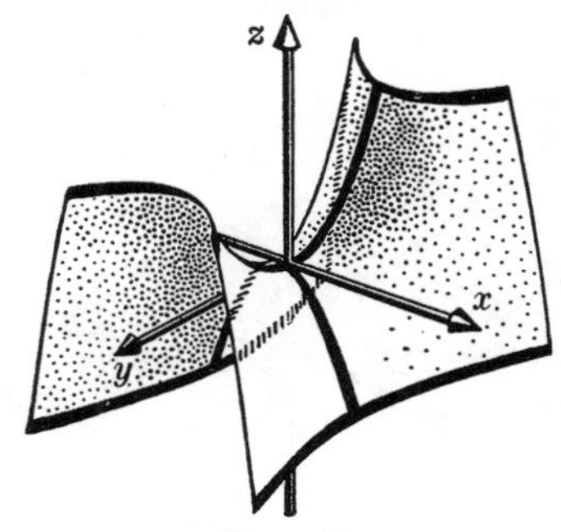

FIG. 43

Equations (44.4) and (44.5) are particular cases of the equation

(45.4) $$ax^2 + by^2 + cz^2 = d.$$

When a, b, c, and d have the same sign, which may be taken as positive in all generality, the intersection of the surface by any one of the planes $x = k$, $y = k$, or $z = k$ is an ellipse, as shown by a process similar to that employed above. In this case the surface is called an *ellipsoid* (see Fig. 44). When a and d are positive and b and c are negative, a plane $x = k$ intersects the surface in an ellipse, real or imaginary according as $ak^2 - d$ is positive or negative; and a plane $y = k$ or $z = k$ intersects the surface in a hyperbola. The surface consists of two sheets, since when $k^2 < \frac{d}{a}$ the plane $x = k$ does not intersect the surface in real points but in an imaginary ellipse. The surface is called a *hyperboloid of two sheets* (see Fig. 45). The first of (44.5) is a special case of this type of surface. When a, b, and d are positive and c is negative, a plane $z = k$ intersects the surface in an ellipse and a plane $x = k$ or $y = k$ intersects it in a hyperbola. The surface is called a *hyperboloid of one sheet* (see Fig. 46). The second of (44.5) is a special case of this type of surface.

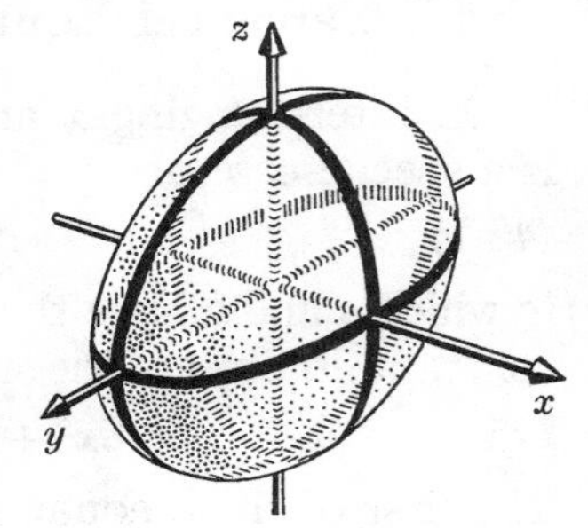

FIG. 44

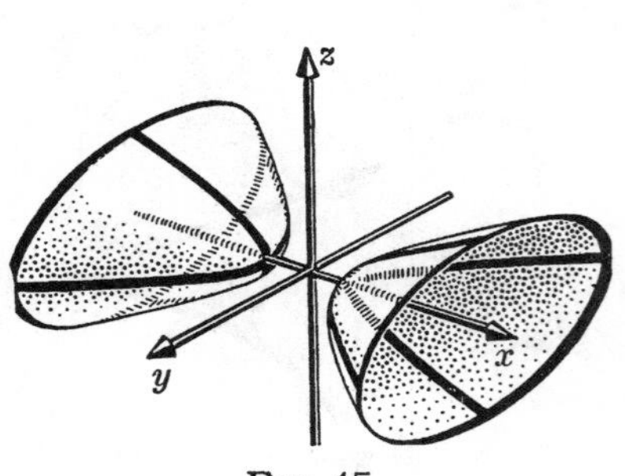

FIG. 45

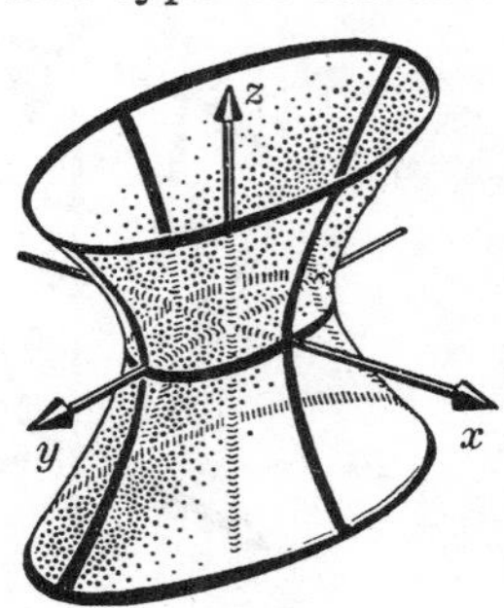

FIG. 46

When d in (45.4) is equal to zero, if (x_1, y_1, z_1) is a point of the surface, so also is (tx_1, ty_1, tz_1) a point of the surface for every value of t. Since all these points lie on the line through the origin and the point (x_1, y_1, z_1), it follows that the surface is a cone with vertex at the origin. The only real solution of equation (45.4) with $d = 0$ is (0, 0, 0) unless one or two of a, b, and c are negative. In order to consider real cones, we may in all generality assume that a and b are positive and c negative. In this case a plane $z = k$ $(\neq 0)$ intersects the cone in an ellipse, and a plane $x = k$ or $y = k$ $(\neq 0)$ in a hyperbola.

The following is a list of canonical, or type, forms of equations of the surfaces just considered:

(45.5) $\dfrac{x^2}{a^2} + \dfrac{y^2}{b^2} = 2\,cz,$ elliptic paraboloid;

(45.6) $\dfrac{x^2}{a^2} - \dfrac{y^2}{b^2} = 2\,cz,$ hyperbolic paraboloid;

(45.7) $\dfrac{x^2}{a^2} + \dfrac{y^2}{b^2} + \dfrac{z^2}{c^2} = d,$ real, imaginary, or point ellipsoid as $d = +1, -1$, or 0;

(45.8) $\dfrac{x^2}{a^2} + \dfrac{y^2}{b^2} - \dfrac{z^2}{c^2} = 1,$ hyperboloid of one sheet;

(45.9) $\dfrac{x^2}{a^2} - \dfrac{y^2}{b^2} - \dfrac{z^2}{c^2} = 1,$ hyperboloid of two sheets;

(45.10) $\dfrac{x^2}{a^2} + \dfrac{y^2}{b^2} - \dfrac{z^2}{c^2} = 0,$ quadric cone.

Other surfaces of the second degree are

(45.11) $\dfrac{x^2}{a^2} + \dfrac{y^2}{b^2} = 1,$ elliptic cylinder;

(45.12) $\dfrac{x^2}{a^2} - \dfrac{y^2}{b^2} = 1,$ hyperbolic cylinder;

(45.13) $x^2 - 4\,ay = 0,$ parabolic cylinder;

(45.14) $\dfrac{x^2}{a^2} - \dfrac{y^2}{b^2} = 0,$ pair of intersecting planes;

(45.15) $x^2 - a^2 = 0,$ pair of parallel planes.

Since all the above equations contain a term in x^2 and none in x, it follows that if (x_1, y_1, z_1) is a point of any one of the surfaces, so also is $(-x_1, y_1, z_1)$, that is, the plane $x = 0$ is a plane of symmetry (see § 14). Any plane of symmetry of a surface is called a *principal plane* of the surface. From equations (45.7) to (45.10) it follows that ellipsoids, hyperboloids, and quadric cones have three principal planes, each two perpendicular to one another; they are the three coordinate planes for these equations. From equations (45.5), (45.6), (45.11), (45.12), and (45.14) it is seen that the paraboloids and elliptic and hyperbolic cylinders have two mutually perpendicular principal planes, as do also two intersecting planes; they are the planes $x = 0$ and $y = 0$ for these equations. Parabolic cylinders and two parallel planes have one principal plane, as follows from (45.13) and (45.15), for which the plane $x = 0$ is the principal plane.

As a result of this discussion it follows that any *quadric surface*, that is, any surface of the second degree, including the *degenerate quadrics* consisting of two planes, has at least one principal plane. It is evident geometrically that any plane through the axis of a quadric of revolution is a principal plane.

For the surfaces with equations (45.7) to (45.10) the origin is a point of symmetry; that is, if (x_1, y_1, z_1) is a point of any one of these surfaces, so also is the point $(-x_1, -y_1, -z_1)$. For an ellipsoid or a hyperboloid the point of symmetry is the mid-point of every chord of the surface through it; such a chord is called a *diameter*. Consequently the ellipsoids and hyperboloids are called the *central quadrics*, the origin being the *center* in the coordinate system used in this section. The point of symmetry of a cone is the *vertex*, and each element of a cone extends indefinitely in both directions.

For elliptic and hyperbolic cylinders with equations (45.11) and (45.12), each point of the z-axis is a point of symmetry. For, if we denote by $P_0(0, 0, z_0)$ a point on the axis, and if x_1, y_1 are solutions of (45.11) or (45.12), then $(x_1, y_1, z_0 + t)$ and $(-x_1, -y_1, z_0 - t)$ are points of the cylinder and are symmetric with respect to P_0 for every value of t. This line of points of symmetry is called the *axis* of the cylinder in each case.

EXERCISES

1. Find the foci of the ellipses in which the ellipsoid

$$6x^2 + 3y^2 + 2z^2 = 6$$

is cut by the coordinate planes.

2. Find an equation of the cone obtained by revolving the line $x - 2y + 1 = 0$, $z = 0$ about the x-axis. Effect a translation of the axes so that the resulting equation of the cone in the $x'y'z'$-system is of the form (45.10).

3. Show that a real ellipsoid (45.7), for which $a < b < c$, is cut by the sphere $x^2 + y^2 + z^2 = b^2$ in two circles.

4. Show that the semi-diameter r with direction cosines λ, μ, ν of an ellipsoid (45.7) is given by

$$\frac{1}{r^2} = \frac{\lambda^2}{a^2} + \frac{\mu^2}{b^2} + \frac{\nu^2}{c^2}$$

and prove that the sum of the squares of the reciprocals of any three mutually perpendicular diameters of an ellipsoid is a constant (see (30.2)).

5. Show that for the surfaces (45.7) to (45.10) the coordinate axes are lines of symmetry (see § 14). Find lines of symmetry of the other quadrics.

6. Show that, with the exception of equations of the central quadrics and of quadric cones, the terms of the second degree in equations (45.5) to (45.15) are the product of two real or imaginary homogeneous factors of the first degree. Does it follow that the same is true of equations of these surfaces in any rectangular coordinate system?

7. Show by means of the argument preceding equation (45.5) that

$$ax^2 + by^2 + cz^2 + 2hxy + 2fxz + 2gyz = 0$$

is an equation of a cone with vertex at the origin, or of two planes through the origin; and that in consequence of Theorem [**41.5**] it is an equation of two planes, if and only if the determinant

$$\begin{vmatrix} a & h & f \\ h & b & g \\ f & g & c \end{vmatrix}$$

is equal to zero.

8. Show that $xy + yz + zx = 0$ is an equation of a cone with vertex at the origin, and find the cosine of an angle between the two lines in which the surface is intersected by the plane $ax + by + cz = 0$,

9. Determine the value of b so that

$$2\,x^2 + by^2 - 12\,z^2 - 3\,xy + 2\,xz + 11\,yz = 0$$

shall be an equation of two planes through the origin (see Ex. 7); and find equations of the planes.

10. Show that an equation of the cone whose vertex is at the center of a real ellipsoid (45.7) and which passes through all the points of intersection of the ellipsoid and the plane $lx + my + nz = 1$ is

$$\frac{x^2}{a^2} + \frac{y^2}{b^2} + \frac{z^2}{c^2} = (lx + my + nz)^2.$$

11. Show that the curve of intersection of the sphere $x^2 + y^2 + z^2 = r^2$ and a real ellipsoid (45.7) for which $a \geq b \geq c$, lies on the surface

$$\left(\frac{1}{a^2} - \frac{1}{r^2}\right)x^2 + \left(\frac{1}{b^2} - \frac{1}{r^2}\right)y^2 + \left(\frac{1}{c^2} - \frac{1}{r^2}\right)z^2 = 0.$$

Determine for what values of r this is an equation of a cone, and for what values it is an equation of two planes (see Ex. 7); and show that when this surface is degenerate and real, the curve of intersection of the ellipsoid and the sphere consists of circles. What is the intersection when this surface is degenerate and imaginary?

46. The Ruled Quadrics

The equation (45.6) of the hyperbolic paraboloid may be written in the form

$$\left(\frac{x}{a} + \frac{y}{b}\right)\left(\frac{x}{a} - \frac{y}{b}\right) = 2\,cz. \tag{46.1}$$

Consider in this connection the two equations

$$\frac{x}{a} + \frac{y}{b} = 2\,ck, \qquad \frac{x}{a} - \frac{y}{b} = \frac{z}{k}. \tag{46.2}$$

For each value of k other than zero these are equations of a line which lies on the paraboloid. In fact, any set of values of x, y, z satisfying (46.2) for a given value of k also satisfies (46.1), as is seen on multiplying together equations (46.2) member by member. And conversely, through each point of the paraboloid (46.1) there passes one of the lines (46.2), the corresponding value of k being obtained from either of equations (46.2) on substituting in the equation the coordinates of the point. Consequently the paraboloid has lying upon it an endless number of lines, or *rulings*. In like manner, the lines whose equations are

$$\frac{x}{a}+\frac{y}{b}=zl, \quad \frac{x}{a}-\frac{y}{b}=\frac{2c}{l} \tag{46.3}$$

for values of l other than zero, lie on the paraboloid (46.1), and through each point of the surface there passes a line of the set (46.3). A surface having an endless number of lines lying upon it, one through each point, is called a *ruled surface*. For example, cones and cylinders are ruled surfaces. Since a hyperbolic paraboloid has two sets of rulings, it is said to be *doubly ruled*.

In like manner, it can be shown that the set of lines whose equations are

$$\frac{x}{a}+\frac{z}{c}=k\left(1+\frac{y}{b}\right), \quad \frac{x}{a}-\frac{z}{c}=\frac{1}{k}\left(1-\frac{y}{b}\right) \tag{46.4}$$

lie on the hyperboloid (45.8), as do also the lines

$$\frac{x}{a}+\frac{z}{c}=l\left(1-\frac{y}{b}\right), \quad \frac{x}{a}-\frac{z}{c}=\frac{1}{l}\left(1+\frac{y}{b}\right). \tag{46.5}$$

Consequently a hyperboloid of one sheet is doubly ruled, a line of each of the sets (46.4) and (46.5) passing through each point of the surface. Hence we have

[46.1] *Hyperbolic paraboloids and hyperboloids of one sheet are doubly ruled surfaces.*

By Theorem **[20.2]** we find that $a, -b, 2k$ are direction numbers of the line (46.2), and $a, b, 2/l$ are direction numbers of the line (46.3). If we denote by k_1 and l_1 the values of k and l for the lines through the point (x_1, y_1, z_1) on the paraboloid (46.1), we have from (46.2) and (46.3)

$$\frac{x_1}{a}+\frac{y_1}{b}=2ck_1, \quad \frac{x_1}{a}-\frac{y_1}{b}=\frac{2c}{l_1}. \tag{46.6}$$

The plane through the point (x_1, y_1, z_1) and containing the two rulings through this point has as an equation (see § 23, Exs. 6 and 7)

$$\begin{vmatrix} x - x_1 & y - y_1 & z - z_1 \\ a & -b & 2k_1 \\ a & b & \dfrac{2}{l_1} \end{vmatrix} = 0,$$

that is,

$$b\left(\frac{1}{l_1} + k_1\right)(x - x_1) + a\left(\frac{1}{l_1} - k_1\right)(y - y_1) - ab(z - z_1) = 0.$$

When the values of k_1 and l_1 given by (46.6) are substituted in this equation, the resulting equation is

$$\frac{x_1}{a^2}(x - x_1) - \frac{y_1}{b^2}(y - y_1) - c(z - z_1) = 0.$$

Since (x_1, y_1, z_1) is a point of the paraboloid (46.1), this equation is reducible to

$$\frac{x_1 x}{a^2} - \frac{y_1 y}{b^2} - c(z + z_1) = 0. \tag{46.7}$$

The direction numbers of any line in this plane are a linear homogeneous combination of direction numbers of the lines (46.2) and (46.3) (see § 20, Ex. 9). Consequently we have as parametric equations of any line through (x_1, y_1, z_1) and contained in the plane (46.7)

$$\begin{aligned} x &= x_1 + (h_1 a + h_2 a)t, \\ y &= y_1 + (-h_1 b + h_2 b)t, \\ z &= z_1 + 2\left(h_1 k_1 + \frac{h_2}{l_1}\right)t, \end{aligned} \tag{46.8}$$

where h_1 and h_2 determine the line, and t is proportional to the distance between the points (x_1, y_1, z_1) and (x, y, z). When these values are substituted in (46.1) in order to find the values of t for the point in which the line (46.8) meets the paraboloid, the resulting equation is reducible, by means of (46.6) and the fact that x_1, y_1, z_1 is a solution of (46.1), to

$$h_1 h_2 t^2 = 0.$$

Thus when h_1 or h_2 is equal to zero, that is, when the line is one of the rulings, every value of t satisfies this equation, meaning that every point of the ruling is a point of the paraboloid. When $h_1 \neq 0$ and $h_2 \neq 0$, the two solutions of the above equation are $t = 0$, that is, each line for any values of h_1 and h_2, both different from zero, meets the surface in a doubly counted point. Consequently every line in the plane (46.7) through the point (x_1, y_1, z_1), other than the rulings, is tangent to the surface at the point; accordingly (46.7) is called the *tangent plane* to the paraboloid at the point (x_1, y_1, z_1) (see Theorem [**48.7**]).

EXERCISES

1. Find equations of the rulings on the paraboloid $x^2 - 2y^2 + 2z = 0$, and in particular of the rulings through the point $(2, -1, -1)$; find also an equation of the tangent plane at this point.

2. Find equations of the rulings on the hyperboloid

$$x^2 + 2y^2 - 4z^2 = 4,$$

and in particular of the rulings through the point (0, 2, 1).

3. Find direction numbers of the rulings (46.4) and (46.5) of a hyperboloid of one sheet (45.8), and show, using equations analogous to (46.8), that an equation of the plane containing the rulings through the point (x_1, y_1, z_1) is

$$\frac{x_1 x}{a^2} + \frac{y_1 y}{b^2} - \frac{z_1 z}{c^2} = 1,$$

which consequently is the tangent plane at (x_1, y_1, z_1).

4. Given the equation

$$\text{(i)} \qquad \frac{x^2}{a^2 - t} + \frac{y^2}{b^2 - t} + \frac{z^2}{c^2 - t} = 1,$$

where $a > b > c$, determine the values of t for which the surface is an ellipsoid, a hyperboloid of one sheet, a hyperboloid of two sheets; discuss in particular the cases $t = b^2$ and $t = c^2$. Show also that each of the coordinate planes intersects the set of quadrics (i) in confocal conics (see § 36, Ex. 14). This set of quadrics is called a set of *confocal quadrics*.

5. Show that through every point in space not on one of the coordinate axes there pass three of the confocal quadrics of Ex. 4, one an ellipsoid, one a hyperboloid of one sheet, and one a hyperboloid of two sheets.

6. Show that
$$\frac{x^2}{a^2} + \frac{y^2}{b^2} - \frac{z^2}{c^2} = k$$
for $k > 0$ is an equation of a family of hyperboloids of one sheet (see § 32); for $k < 0$, an equation of a family of hyperboloids of two sheets; for $k = 0$, an equation of a quadric cone. In what sense is this cone the *asymptotic cone* of the two families of hyperboloids?

7. Show that for a suitable value of m in terms of k each of the planes

(i) $$\frac{y}{b} + \frac{z}{c} = m\left(1 - \frac{x}{a}\right), \quad \frac{y}{b} - \frac{z}{c} = \frac{1}{m}\left(1 + \frac{x}{a}\right)$$

passes through a ruling (46.4), and consequently equations (i) are equations of the ruling. Show also that
$$\frac{y}{b} + \frac{z}{c} = n\left(1 + \frac{x}{a}\right), \quad \frac{y}{b} - \frac{z}{c} = \frac{1}{n}\left(1 - \frac{x}{a}\right)$$
are equations of the rulings (46.5).

47. Quadrics Whose Principal Planes Are Parallel to the Coordinate Planes

Each of equations (45.5) to (45.15) is a special case of an equation

(47.1) $$ax^2 + by^2 + cz^2 + 2\,lx + 2\,my + 2\,nz + d = 0,$$

with the understanding that a, b, c in this equation are not the same as in equations (45.5) to (45.15). For example, if $c = l = m = d = 0$ and we replace a by $1/a^2$, b by $1/b^2$, and n by $-c$, we have equation (45.5). Now we shall prove that (47.1), whatever be the coefficients provided only that a, b, and c are not all zero, is an equation of a quadric or degenerate quadric, by showing that by a suitable transformation of coordinates equation (47.1) can be transformed into one of the forms (45.5) to (45.15). We consider the three cases, when a, b, c are different from zero, when one of a, b, c is equal to zero, and when two of them are equal to zero.

Case 1. **$a \neq 0$, $b \neq 0$, $c \neq 0$.** On completing the squares in x, y, and z in (47.1), we have
$$a\left(x + \frac{l}{a}\right)^2 + b\left(y + \frac{m}{b}\right)^2 + c\left(z + \frac{n}{c}\right)^2 = k,$$

where
$$k = \frac{l^2}{a} + \frac{m^2}{b} + \frac{n^2}{c} - d.$$

If then we effect the translation of axes defined by

$$(47.2) \qquad x' = x + \frac{l}{a}, \quad y' = y + \frac{m}{b}, \quad z' = z + \frac{n}{c},$$

we obtain the equation

$$(47.3) \qquad ax'^2 + by'^2 + cz'^2 = k.$$

If $k \neq 0$ and a, b, c have the same sign, the surface is a real or imaginary ellipsoid. If $k \neq 0$ and two of the numbers a/k, b/k, c/k are positive and one is negative, equation (47.3) is either of the form (45.8) or one of the forms

$$-\frac{x^2}{a^2} + \frac{y^2}{b^2} + \frac{z^2}{c^2} = 1, \quad \frac{x^2}{a^2} - \frac{y^2}{b^2} + \frac{z^2}{c^2} = 1.$$

Consequently the surface is a hyperboloid of one sheet. If $k \neq 0$ and one of the numbers a/k, b/k, c/k is positive and the other two negative, equation (47.3) is either of the form (45.9) or one of the forms

$$-\frac{x^2}{a^2} + \frac{y^2}{b^2} - \frac{z^2}{c^2} = 1, \quad -\frac{x^2}{a^2} - \frac{y^2}{b^2} + \frac{z^2}{c^2} = 1.$$

Consequently the surface is a hyperboloid of two sheets. In every case, as follows from (47.2), the center of the surface is at the point $(-l/a, -m/b, -n/c)$.

If $k = 0$ and a, b, c have the same sign, the surface is a point ellipsoid (an imaginary cone). If $k = 0$ and one or two of a, b, c are positive and the others negative, the surface is a quadric cone with the vertex at the point $(-l/a, -m/b, -n/c)$.

Case 2. $a \neq 0$, $b \neq 0$, $c = 0$. On completing the squares in x and y, we have

$$a\left(x + \frac{l}{a}\right)^2 + b\left(y + \frac{m}{b}\right)^2 + 2nz + k = 0,$$

where
$$k = d - \frac{l^2}{a} - \frac{m^2}{b}.$$

If $n \neq 0$ and we effect the transformation defined by the first two of equations (47.2) and $z' = z + k/2n$, we obtain

$$ax'^2 + by'^2 + 2nz' = 0,$$

which is of the form (45.5) or (45.6) according as a and b have the same or opposite signs. Consequently the surface is a paraboloid.

If $n = 0$ and we effect the transformation defined by the first two of equations (47.2) and $z' = z$, we obtain

$$ax'^2 + by'^2 + k = 0.$$

Comparing this equation with equations (45.11), (45.12), and (45.14), we have that the surface is an elliptic or hyperbolic cylinder if $k \neq 0$, and two intersecting planes, real or imaginary, if $k = 0$.

If one of the coefficients a, b, c other than c is equal to zero, on using the above methods we get an equation obtained from one of the equations (45.5), (45.6), (45.11), (45.12), and (45.14) on interchanging x and z, or y and z, as the case may be, with the corresponding interpretation.

Case 3. $\boldsymbol{a \neq 0,\ b = c = 0.}$ When the equation is written in the form

$$a\left(x + \frac{l}{a}\right)^2 + 2\,my + 2\,nz + k = 0, \tag{47.4}$$

where

$$k = d - \frac{l^2}{a},$$

if not both m and n are zero and we effect the transformation defined by

$$x' = x + \frac{l}{a}, \qquad y' = \frac{my + nz + \dfrac{k}{2}}{\sqrt{m^2 + n^2}}, \qquad z' = \frac{ny - mz}{\sqrt{m^2 + n^2}},$$

the resulting equation is

$$ax'^2 + 2\sqrt{m^2 + n^2}\, y' = 0. \tag{47.5}$$

Consequently the surface is a parabolic cylinder.

If $m = n = 0$, it follows from (47.4) that the surface is degenerate, consisting of two parallel, real or imaginary, planes if $k \neq 0$, and of two coincident planes if $k = 0$.

When any two of the coefficients a, b, c are equal to zero, similar results follow. Hence we have

[47.1] *An equation $ax^2 + by^2 + cz^2 + 2\,lx + 2\,my + 2\,nz + d = 0$, in which not all the coefficients a, b, c are zero, is an equation of a quadric, which may be degenerate; its principal plane or planes are parallel to the coordinate planes.*

The last part of this theorem follows from the fact that in each case equation (47.1) was transformed by a translation (30.1) into one of the forms (45.5) to (45.15), except in Case 3, where, however, the $y'z'$-plane is parallel to the yz-plane.

EXERCISES

1. Show that a surface with an equation

$$a(x - x_0)^2 + b(y - y_0)^2 + c(z - z_0)^2 = 1,$$

in which a, b, c are positive, is an ellipsoid with the point (x_0, y_0, z_0) as center; discuss also the cases when a, b, c do not all have the same sign, and find equations of the principal planes.

2. Show that a surface with an equation

$$a(x - x_0)^2 + b(y - y_0)^2 + c(z - z_0) = 0,$$

in which $a, b, c \neq 0$ is an elliptic or hyperbolic paraboloid according as a and b have the same or different signs. Find equations of the principal planes. What is the surface when $c = 0$?

3. Identify the surface defined by each of the following equations, and determine its position relative to the coordinate axes:

a. $9x^2 + 4y^2 + 36z^2 - 36x + 9y + 4 = 0.$
b. $x^2 - 4y^2 - z^2 - 4x - 24y + 4z - 32 = 0.$
c. $x^2 + 4y^2 + 4x - 8y - 6z + 14 = 0.$
d. $x^2 - 4z^2 + 5y - x + 8z = 0.$
e. $y^2 - 4z^2 + 4y + 4z + 3 = 0.$

4. For what values of a, b, c, and d is the surface

$$ax^2 + by^2 + cz^2 + d = 0 \qquad \text{(i)}$$

a real ellipsoid; for what values a hyperboloid of one sheet; for what values a hyperboloid of two sheets?

5. Find the projection upon each of the coordinate planes of the intersection of $x^2 - 2y^2 - 3z^2 - 6 = 0$ and $x - 3y + 2z = 0$ (see § 44, Ex. 11). Of what type is the curve of intersection?

6. Find an equation of the projection upon each of the coordinate planes of the intersection of the central quadric (i) of Ex. 4 and the plane $lx + my + nz = 0$, and identify each of these curves.

7. Show that, if the line with equations

(i) $$x = x_1 + ut, \quad y = y_1 + vt, \quad z = z_1 + wt$$

meets a central quadric with equation (i) of Ex. 4 in two points, the point (x_1, y_1, z_1) is the mid-point of the intercepted chord if it lies in the plane

(ii) $$aux + bvy + cwz = 0.$$

Why is this plane appropriately called a *diametral plane* of the quadric, and under what conditions is it perpendicular to the chords which it bisects?

8. Show that if (x_1, y_1, z_1) is a point on the intersection of the plane (ii) of Ex. 7 and the quadric (i) of Ex. 4, equations (i) of Ex. 7 are equations of a tangent to the quadric at the point (x_1, y_1, z_1).

9. Find an equation of the plane which is the locus of the mid-points of the chords of the elliptic paraboloid $x^2 + 3y^2 = 2z$ whose direction numbers are 1, 2, 3.

10. Let P_1 and P_2 be two points on a real ellipsoid (45.7) such that P_2 is on the diametral plane of chords parallel to the line OP_1, where O is the center (0, 0, 0); prove that P_1 is on the diametral plane of chords parallel to OP_2 (see Ex. 7).

48. The General Equation of the Second Degree in *x*, *y*, and *z*. The Characteristic Equation. Tangent Planes to a Quadric

The most general equation of the second degree in x, y, and z is of the form

$$(48.1) \quad ax^2 + by^2 + cz^2 + 2hxy + 2fxz + 2gyz + 2lx + 2my + 2nz + d = 0.$$

We shall show that any such equation, for which not all the coefficients of the terms of the second degree are zero, is an equation of a quadric. We prove this by showing that in every case there exists a transformation of coordinates by which the equation is transformed into an equation of the form (47.1), which was shown in § 47 to be an equation of a quadric. In § 45 we observed that any one of the surfaces (45.5) to (45.15) has at least one plane of symmetry, and in Theorem [**47.1**] that the plane or planes of symmetry for a quadric with an equation (47.1) are parallel to the coordinate planes. Accordingly we seek the planes of symmetry of the surface with an equation (48.1).

The equations

$$(48.2) \qquad x = x_1 + ut, \qquad y = y_1 + vt, \qquad z = z_1 + wt$$

are parametric equations of the line through the point (x_1, y_1, z_1), and with direction numbers u, v, w. If we substitute these expressions for x, y, and z in (48.1) and collect the terms in t^2 and t, we have

$$(48.3) \quad (au^2 + bv^2 + cw^2 + 2\,huv + 2\,fuw + 2\,gvw)t^2 \\ + 2[(ax_1 + hy_1 + fz_1 + l)u + (hx_1 + by_1 + gz_1 + m)v \\ + (fx_1 + gy_1 + cz_1 + n)w]t + F(x_1, y_1, z_1) = 0,$$

where $F(x_1, y_1, z_1)$ denotes the value of the left-hand member of (48.1) when x, y, and z are replaced by x_1, y_1, and z_1 respectively. For fixed values of x_1, y_1, z_1; u, v, w equation (48.3) is a quadratic equation in t, whose roots are proportional to the distances from (x_1, y_1, z_1) to the two points of intersection, if any, of the surface (48.1) and the line with equations (48.2). In order that (x_1, y_1, z_1) shall be the mid-point of the segment between these two points, the roots of this equation must differ only in sign; hence the coefficient of t must be equal to zero, that is,

$$(48.4) \quad (ax_1 + hy_1 + fz_1 + l)u + (hx_1 + by_1 + gz_1 + m)v \\ + (fx_1 + gy_1 + cz_1 + n)w = 0.$$

As a first consequence of this equation we prove the theorem

[**48.1**] *The mid-points of any set of parallel chords of the surface with an equation* (48.1) *lie in a plane.*

In fact, since the quantities u, v, w have constant values for a set of parallel chords, on reassembling the terms in (48.4), we have that the coordinates of the mid-point of each chord satisfy the equation

$$(48.5) \quad (au + hv + fw)x + (hu + bv + gw)y \\ + (fu + gv + cw)z + (lu + mv + nw) = 0,$$

which evidently is an equation of a plane.

If this plane is to be a plane of symmetry (see § 45) of the surface (48.1), u, v, and w must be such that the chords are normal to the plane. By Theorem [17.5] the coefficients of x, y, and z in (48.5) are direction numbers of any normal to the plane. Accordingly, if the chords are to be normal to the plane, we must have

$$\text{(48.6)} \qquad \frac{au + hv + fw}{u} = \frac{hu + bv + gw}{v} = \frac{fu + gv + cw}{w} = r,$$

when r denotes the common value of these ratios. When we write these equations in the form

$$\text{(48.7)} \qquad \begin{aligned} (a - r)u + hv + fw &= 0, \\ hu + (b - r)v + gw &= 0, \\ fu + gv + (c - r)w &= 0, \end{aligned}$$

we have by Theorem [22.1] that they admit a common solution u, v, w, not all zero, if and only if r is such that

$$\text{(48.8)} \qquad \begin{vmatrix} a - r & h & f \\ h & b - r & g \\ f & g & c - r \end{vmatrix} = 0.$$

Thus r must be a solution of the cubic equation

$$\text{(48.9)} \qquad r^3 - Ir^2 + Jr - D = 0,$$

the coefficients I, J, and D being defined by

$$\text{(48.10)} \qquad \begin{aligned} I &= a + b + c, \\ J &= bc + ca + ab - f^2 - g^2 - h^2, \\ D &= \begin{vmatrix} a & h & f \\ h & b & g \\ f & g & c \end{vmatrix}. \end{aligned}$$

Equation (48.8) is called the *characteristic equation* of equation (48.1). We now derive properties of the characteristic equation by means of which we shall prove that (48.1) is an equation of a quadric.

Let r_1 and r_2 be two different roots of this equation, and denote by u_1, v_1, w_1 and u_2, v_2, w_2 solutions of equations (48.7) as r takes the values r_1 and r_2 respectively. If in equations (48.7) we replace u, v, w, and r by u_1, v_1, w_1, and r_1 respectively and form the sum of these equations after multiplying them by u_2, v_2, and w_2 respectively, we obtain

$$au_1u_2 + bv_1v_2 + cw_1w_2 + h(u_2v_1 + u_1v_2) + f(u_2w_1 + u_1w_2) \\ + g(v_2w_1 + v_1w_2) - r_1(u_1u_2 + v_1v_2 + w_1w_2) = 0.$$

If, in similar manner, we replace u, v, w, and r in (48.7) by u_2, v_2, w_2, and r_2 and form the sum of these equations after multiplying them by u_1, v_1, and w_1 respectively, we obtain the above equation with the exception that r_1 is replaced by r_2. Having obtained these two equations, if we subtract one from the other and note that by hypothesis $r_1 \neq r_2$, we have

$$u_1u_2 + v_1v_2 + w_1w_2 = 0. \tag{48.11}$$

With the aid of this result we shall establish the important theorem

[48.2] *The roots of the characteristic equation of an equation of the second degree whose coefficients are real are all real numbers.*

For, suppose one of the roots were imaginary, say $r_1 = \sigma + i\tau$, where σ and τ are real numbers and $i = \sqrt{-1}$. From the theory of algebraic equations it follows that $r_2 = \sigma - i\tau$ also is a root. If we replace r in (48.7) by r_1 and solve the resulting equations for u_1, v_1, and w_1 by means of Theorem [22.1], we have that at least one of these quantities is imaginary; for, if in (48.7) everything is real but $r = \sigma + i\tau$ and we equate to zero the real and imaginary parts of each equation, we have that $\tau = 0$ unless $u = v = w = 0$. Accordingly we put

$$u_1 = \alpha_1 + i\alpha_2, \qquad v_1 = \beta_1 + i\beta_2, \qquad w_1 = \gamma_1 + i\gamma_2,$$

where the α's, β's, and γ's are real numbers. When r in (48.7) is replaced by $\sigma - i\tau$, a solution of the resulting equations is

$$u_2 = \alpha_1 - i\alpha_2, \qquad v_2 = \beta_1 - i\beta_2, \qquad w_2 = \gamma_1 - i\gamma_2,$$

since the set of equations (48.7) for r_2 is obtained from the above set on changing i to $-i$. For these two sets of solutions it follows from (48.11) that

$$\alpha_1^2 + \beta_1^2 + \gamma_1^2 + \alpha_2^2 + \beta_2^2 + \gamma_2^2 = 0,$$

which evidently is impossible, since all the quantities in this equation are real and not all zero. Hence Theorem [**48.2**] is proved. As a consequence of this result and equation (48.11) we have

[**48.3**] *Planes of symmetry of the surface* (48.1) *corresponding to two unequal roots of the characteristic equation are perpendicular to one another.*

Another theorem which we need in the proof that every equation of the second degree is an equation of a quadric is the following:

[**48.4**] *Not all the roots of the characteristic equation of an equation of the second degree with real coefficients are equal to zero.*

In order to prove it, we observe that if all the roots are zero, equation (48.9) must be $r^3 = 0$, that is, $I = J = D = 0$. In this case $I^2 - 2J = 0$, and from (48.10) we have

$$I^2 - 2J = a^2 + b^2 + c^2 + 2f^2 + 2g^2 + 2h^2.$$

This expression can be zero only in case a, b, c, f, g, h are all zero, in which case equation (48.1) is not of the second degree; and the theorem is proved.

From Theorems [**48.2**] and [**48.4**] it follows that at least one of the roots of the characteristic equation is a real number different from zero. For this root equations (48.7) admit at least one solution u, v, w, not all zero, since (48.8) is satisfied. If then the principal plane having equation (48.5) with these values of u, v, and w is taken as the plane $z' = 0$ of a new rectangular coordinate system, the transform of equation (48.1) does not involve terms in $x'z'$, $y'z'$, or z', since, if (x_1', y_1', z_1') is on the surface, then $(x_1', y_1', -z_1')$ is also. Hence the equation is of the form

$$(48.12) \qquad a'x'^2 + b'y'^2 + c'z'^2 + 2h'x'y' + 2l'x' + 2m'y' + d' = 0,$$

where some of the coefficients may be equal to zero, but not all those of the terms of the second degree. If $h' = 0$, (48.12) is of the form (47.1). If $h' \neq 0$ and we effect the rotation of the axes about the z'-axis defined by

$$x' = x'' \cos\theta - y'' \sin\theta, \quad y' = x'' \sin\theta + y'' \cos\theta, \quad z' = z'',$$

where θ is a solution of the equation (see § 40)

$$h' \tan^2\theta + (a' - b') \tan\theta - h' = 0,$$

the resulting equation in x'', y'', and z'' is of the form (47.1). Hence we have established the theorem

[48.5] *Any equation of the second degree in x, y, and z is an equation of a quadric, which may be degenerate.*

As a consequence of the preceding results we have

[48.6] *When any plane cuts a quadric, the curve of intersection is a conic, which may be degenerate.*

In fact, given any plane, in a suitably chosen coordinate system it is the plane $z = 0$. Solving this equation simultaneously with (48.1), we have an equation of the second degree in x and y which is an equation of a conic, as shown in § 40.

We now make use of equation (48.3) to derive an equation of the tangent plane to a quadric (48.1) at a point (x_1, y_1, z_1), proceeding as was done in § 46 for a hyperbolic paraboloid with equation (45.6). When (x_1, y_1, z_1) is a point of the quadric (48.1), $F(x_1, y_1, z_1) = 0$ in (48.3). In order that the line (48.2) shall be tangent to the quadric, $t = 0$ must be a double root of (48.3). Consequently we must have equation (48.4) satisfied by u, v, and w; and for any set of values of u, v, and w satisfying (48.4) the corresponding equations (48.2) are equations of a tangent line. This means that any line through (x_1, y_1, z_1) perpendicular to the line through this point and with direction numbers

$$\begin{aligned} &ax_1 + hy_1 + fz_1 + l, \\ &hx_1 + by_1 + gz_1 + m, \qquad (48.13) \\ &fx_1 + gy_1 + cz_1 + n, \end{aligned}$$

is tangent to the quadric at (x_1, y_1, z_1). From geometric considerations we have that all these tangents lie in a plane, called the *tangent plane* to the quadric. We prove this analytically by observing that if (x, y, z) is any point on one of these lines, $x - x_1$, $y - y_1$, $z - z_1$ are direction numbers of the line. Hence we have

$$(ax_1 + hy_1 + fz_1 + l)(x - x_1) + (hx_1 + by_1 + gz_1 + m)(y - y_1) + (fx_1 + gy_1 + cz_1 + n)(z - z_1) = 0,$$

which is an equation of a plane.

On multiplying out the terms in this equation and making use of the fact that x_1, y_1, z_1 is a solution of (48.1), we have

[48.7] *An equation of the tangent plane to the quadric* (48.1) *at the point* (x_1, y_1, z_1) *is*

$$\begin{aligned}(48.14)\qquad & ax_1x + by_1y + cz_1z + h(x_1y + y_1x) + f(x_1z + z_1x)\\ & + g(y_1z + z_1y) + l(x + x_1) + m(y + y_1)\\ & + n(z + z_1) + d = 0.\end{aligned}$$

Since (48.13) are direction numbers of the normal to the tangent plane at the point (x_1, y_1, z_1), called the *normal to the quadric* at this point, this normal has equations

$$(48.15)\qquad \frac{x - x_1}{ax_1 + hy_1 + fz_1 + l} = \frac{y - y_1}{hx_1 + by_1 + gz_1 + m} = \frac{z - z_1}{fx_1 + gy_1 + cz_1 + n}.$$

EXERCISES

1. Find the roots of the characteristic equation of the quadric

$$2x^2 + 5y^2 + 3z^2 + 4xy - 3x + 4y - 6z - 3 = 0;$$

determine the corresponding planes of symmetry, and a system of coordinates in terms of which the equation has one of the canonical forms of § 45.

2. Show that when an equation of a quadric does not involve terms in xy, xz, and yz, the roots of the characteristic equation are the coefficients of the second-degree terms. What are the solutions of equations (48.7) in this case?

3. Show that the characteristic equation is unaltered by a translation of the axes, that is, by a transformation to parallel axes.

4. Prove that the normals to a central quadric $ax^2 + by^2 + cz^2 = 1$ at all the points in which the quadric is met by a plane parallel to one of the coordinate planes intersect two fixed lines, one in each of the other coordinate planes and parallel to the intersecting plane.

5. Denote by Q the point in which the normal to a real ellipsoid (45.7) at a point P meets the plane $z = 0$. Find the locus of the mid-point of PQ.

6. Find equations of the tangent plane and of the normal at the point $(1, 1, -2)$ to the quadric

$$x^2 + y^2 - z^2 + 2\,xy + xz + 4\,yz - x + y + z + 4 = 0.$$

7. Prove that the line $x - 2 = 0$, $z - 1 = 0$ lies entirely on the quadric $xy + 3\,xz - 2\,yz - 3\,x - 6\,z + 6 = 0$. Find an equation of the tangent plane to the quadric at the point of this line for which $y = y_1$ and show that the line lies in the tangent plane, and that as y_1 takes different values one gets a family of planes through the line. May one conclude from this that the quadric is neither a cone nor a cylinder?

8. Find the conditions to be satisfied by u, v, w so that the line (48.2) is a ruling of the quadric (48.1) through the point (x_1, y_1, z_1), that is, lies entirely on the quadric; and show that when these conditions are satisfied, the line lies entirely in the tangent plane to the quadric at the point (x_1, y_1, z_1).

9. Show that any of the quadrics (45.5) to (45.15) is cut by a set of parallel planes in similar conics and that the principal axes of any two of these conics are parallel. Is this true of a quadric with equation (48.1)?

10. Show that the normals at P to the three confocal quadrics through a point P are mutually perpendicular (see § 46, Ex. 4).

11. The plane (48.14) is called the *polar plane* of the point $P_1(x_1, y_1, z_1)$ with respect to the quadric (48.1), whether P_1 is on the quadric or not; and P_1 is called the *pole* of the plane. Show that if the polar plane intersects the quadric and P_2 is any point on the intersection, the tangent plane to the quadric at P_2 passes through P_1, and the line through P_1 and P_2 is tangent to the quadric at P_2; also that the locus of all such lines through P_1 is a quadric cone.

12. Show, with the aid of (48.3), that if an endless number of tangents can be drawn to the quadric (48.1) from a point $P_1(x_1, y_1, z_1)$ not on the quadric, the points of contact lie in the polar plane of the point P_1. Where do such points P_1 lie relative to each type of quadric?

13. Find the pole of the plane $3x - 2y - 3z - 1 = 0$ with respect to the quadric $x^2 + y^2 + z^2 + 4xy - 2xz + 4yz - 1 = 0$.

14. Show that the poles with respect to the quadric (48.1) of the planes through a line are collinear.

15. Show that if $P_1(x_1, y_1, z_1)$ is a point from which tangents cannot be drawn to a quadric (48.1), the poles of three planes through P_1 and not having a line in common determine the polar plane of P_1 (see § 34, Ex. 13).

49. Centers. Vertices. Points of Symmetry

Having shown that any equation of the second degree in x, y, and z is an equation of a quadric, we shall establish in this section and the next criteria which enable one to determine the particular type of quadric defined by a given equation.

As a step in this direction we seek the conditions upon the coefficients of equation (48.1) in order that the quadric have a *center*, that is, that there be a point (x_1, y_1, z_1) not on the quadric which is the mid-point of every chord (48.2) through the point, whatever be u, v, and w. Referring to equation (48.4) and the discussion leading up to it, we see that x_1, y_1, z_1 must be a common solution of the equations

$$\begin{aligned} ax + hy + fz + l &= 0, \\ hx + by + gz + m &= 0, \\ fx + gy + cz + n &= 0. \end{aligned} \tag{49.1}$$

In accordance with Theorem [21.8] these equations have one and only one common solution when their determinant D, defined by (48.10), is different from zero. Since the property that a point is a center is independent of the coordinate system used, we may consider the case when the coordinate system is such that $f = g = h = 0$. In this case $D = abc$, that is, the product of the coefficients of x^2, y^2, and z^2. Applying the above test to equations (45.5) to (45.15), we see that ellipsoids, hyperboloids, and cones are the only quadrics for which $D \neq 0$; and in the coordinate system of these equations the point whose coordinates satisfy (49.1) is the origin. In the case of a cone the solution of (49.1) is a point of the cone, the *vertex*, so that ellipsoids and hyperboloids are the only quadrics which have a center.

In order to discuss the case $D \neq 0$ more fully, and also the case $D = 0$, we denote by $F(x, y, z)$ the expression which is the left-hand member of equation (48.1), and note that it may be written

$$(49.2) \qquad F(x, y, z) = (ax + hy + fz + l)x + (hx + by + gz + m)y \\ + (fx + gy + cz + n)z + lx + my + nz + d.$$

If x_1, y_1, z_1 is a solution of equations (49.1), and the point is on the quadric (48.1), it follows from (49.2) that x_1, y_1, z_1 is a solution also of the equation

$$(49.3) \qquad lx + my + nz + d \doteq 0.$$

From Theorem [**27.2**] for the case in which $w = 1$ it follows that this equation and equations (49.1) have a common solution, if and only if $\Delta = 0$, where by definition

$$(49.4) \qquad \Delta = \begin{vmatrix} a & h & f & l \\ h & b & g & m \\ f & g & c & n \\ l & m & n & d \end{vmatrix}.$$

Δ so defined is called the *discriminant* of $F(x, y, z)$. From the foregoing results we have the theorem

[49.1] *When $D \neq 0$ for an equation* (48.1), *the latter is an equation of an ellipsoid or a hyperboloid when $\Delta \neq 0$, and of a cone when $\Delta = 0$. In either case the point whose coordinates are the unique solution of equations* (49.1) *is a point of symmetry of the quadric.*

The latter part of this theorem follows from the fact that if for a value of t the values of x, y, and z given by (48.2) are coordinates of a point on the quadric, so also are the values when t is replaced by $-t$, as follows from (48.3) and (49.1).

Applying the method of § 21 (p. 110) to equations (49.1), we have in place of equations (21.7), (21.8), and (21.9)

$$Dx - L = 0, \quad Dy - M = 0, \quad Dz - N = 0,$$

where L, M, and N are the cofactors of l, m, and n respectively in Δ; and D is seen to be the cofactor of d. Hence if $D = 0$, equations (49.1) do not admit a common solution unless

$$L = M = N = 0, \tag{49.5}$$

in which case $\Delta = 0$. In accordance with the discussion in § 24, when $D = 0$ and equations (49.1) admit a common solution, the planes defined by (49.1) either meet in a line or are coincident; that is, there is a line or a plane all of whose points are points of symmetry of the quadric. Accordingly we have

[49.2] *When* $D = 0$ *for an equation* (48.1) *of a quadric, and equations* (49.1) *admit a common solution,* $\Delta = 0$ *also; and the quadric so defined has a line or plane of points of symmetry.*

Since in either case there is at least one line of points of symmetry, on choosing an $x'y'z'$-system of coordinates in which this line is the z'-axis, we have that the corresponding equations (49.1) must be satisfied by $x' = 0$, $y' = 0$, and any value of z', and consequently

$$c' = f' = g' = l' = m' = n' = 0.$$

If $h' \neq 0$ in the equation of the quadric, by a suitable rotation of the axes about the z'-axis we obtain a coordinate system in terms of which an equation of the quadric is

$$a'x'^2 + b'y'^2 + d' = 0.$$

Comparing this equation with equations (45.11), (45.12), (45.14), and (45.15), we have the theorem

[49.3] *When* $D = 0$ *for an equation* (48.1) *of a quadric, and equations* (49.1) *have a common solution, the quadric is an elliptic or hyperbolic cylinder, or is degenerate in the case when this common solution satisfies equation* (49.3).

All types of quadrics except paraboloids and parabolic cylinders are covered by Theorems [**49.1**] and [**49.3**]. Hence we have

[**49.4**] *Paraboloids and parabolic cylinders do not have any points of symmetry.*

From the fact that the above theorems discriminate between geometric properties of the quadrics, it follows that if D is not or is equal to zero in one coordinate system, the same is true for any coordinate system; the same observation applies to Δ. In the next section we shall show that D and Δ are invariants under any transformation of rectangular coordinates.

50. The Invariants I, J, D, and Δ

In this section we show that I, J, D, and Δ, as defined by (48.10) and (49.4), are invariants under any change of coordinates, in the sense that when a general transformation of coordinates (30.10) is applied to equation (48.1) each of the above functions of the coefficients in (48.1) is equal to the same function of the coefficients of the equation in the new coordinates.

We denote by $f(x, y, z)$ the terms of the second degree in (48.1) and consider the expression

$$f(x, y, z) - r(x^2 + y^2 + z^2), \tag{50.1}$$

that is,

$$(a - r)x^2 + (b - r)y^2 + (c - r)z^2 + 2\,hxy + 2fxz + 2\,gyz. \tag{50.2}$$

From Theorem [**41.5**], which applies also to a homogeneous expression of the second degree in three unknowns, it follows that the equation

$$\begin{vmatrix} a - r & h & f \\ h & b - r & g \\ f & g & c - r \end{vmatrix} = 0 \tag{50.3}$$

is the condition that the expression (50.2) be the product of two linear homogeneous factors; that is, for each root of the cubic equation (50.3), written in the form

$$(50.4) \qquad r^3 - Ir^2 + Jr - D = 0,$$

the equation obtained by equating (50.2) to zero is the equation of two planes through the origin.

When a transformation (30.11) is applied to (50.2), the latter is transformed into

$$(50.5) \qquad f'(x', y', z') - r(x'^2 + y'^2 + z'^2),$$

where $f'(x', y', z')$ is the transform of $f(x, y, z)$; that is, (50.5) is

$$(a' - r)x'^2 + (b' - r)y'^2 + (c' - r)z'^2 + 2\,h'x'y' + 2f'x'z' + 2\,g'y'z'.$$

If (50.2) is the product of two homogeneous factors of the first degree in x, y, z, each of these factors is transformed by a transformation (30.11) into a factor of the same kind, that is, (50.5) is the product of two such factors, and consequently we have, analogously to (50.4),

$$r^3 - I'r^2 + J'r - D' = 0,$$

where I', J', and D' are the same functions of a', b', $\cdots$, h' as I, J, and D are of the corresponding coefficients without primes. Since r is unaltered by the transformation, this equation and (50.4) must have the same roots, and it follows that $I = I'$, $J = J'$, $D = D'$. Since a general transformation of axes may be obtained by applying first a translation (30.1) and then a rotation (30.11), and since the former transformation does not affect the coefficients of the second-degree terms in (48.1), we have established the theorem

[50.1] *The functions I, J, and D of the coefficients of the second-degree terms of an equation of the second degree in three unknowns are invariants under any change of rectangular coordinate axes.*

We apply this theorem to the consideration of the case when two of the roots of the characteristic equation (50.4) are equal and not zero, that is, one of the roots, say r_1, is a double root, and $r_1 \neq 0$. Considering it first as merely a root of the characteristic equation, we find a solution of equations (48.7), and take the principal plane so determined for the plane $x' = 0$ of a new rectangular coordinate system. Whatever be the other coordinate planes, in the transformed equation we have $h' = f' = 0$, so that the new characteristic equation is

$$\begin{vmatrix} a' - r & 0 & 0 \\ 0 & b' - r & g' \\ 0 & g' & c' - r \end{vmatrix} = 0. \tag{50.6}$$

In this coordinate system equations (48.7) are

$$\begin{gathered} (a' - r)u + 0\,v + 0\,w = 0, \qquad 0\,u + (b' - r)v + g'w = 0, \\ 0\,u + g'v + (c' - r)w = 0. \end{gathered} \tag{50.7}$$

Since 1, 0, 0 are direction numbers of any normal to the plane $x' = 0$, for this solution of (50.7) $r = a'$. If we denote this root by r_1, the first of (50.7) is satisfied by any values of u, v, and w, and the last two by 0, g', $r_1 - b'$, provided that

$$(b' - r_1)(c' - r_1) - g'^2 = 0.$$

This is the condition that r_1 $(= a')$ be a double root of (50.6), as is seen when (50.6) is written in the form

$$(a' - r)[(b' - r)(c' - r) - g'^2] = 0.$$

If a plane whose normals have direction numbers 0, g', $r_1 - b'$ is taken for the plane $y' = 0$, without changing the coordinate x', in this new $x'y'z'$-coordinate system, equations (50.7) must be satisfied by 0, 1, 0, which are direction numbers of any normal to the plane $y' = 0$, and consequently in this system $b' = r_1$, $g' = 0$, and $c' = r_3$. If then any plane perpendicular to the planes $x' = 0$ and $y' = 0$ is taken for the plane $z' = 0$, in this final coordinate system the transform of (48.1) is

$$r_1(x'^2 + y'^2) + r_3 z'^2 + 2\,n'z' + d' = 0,$$

there being no terms of the first degree in x' and y', since the planes $x' = 0$ and $y' = 0$ are planes of symmetry. Any plane $z' =$ const. which meets the quadric intersects it in a circle (which may be a point circle or an imaginary circle). Consequently the quadric is a surface of revolution. Since the geometric character of the surface is independent of the coordinate system, we have the theorem

[50.2] *When a nonzero root of the characteristic equation* (48.8) *of an equation* (48.1) *is a double root, the quadric is a surface of revolution; for this root equations* (48.7) *admit an endless number of solutions defining principal planes, all passing through the axis of the surface.*

We proceed now to the proof that the discriminant Δ of $F(x, y, z)$, that is, of the left-hand member of (48.1), is an invariant. To this end we consider the discriminant of the equation

$$F(x, y, z) - r(x^2 + y^2 + z^2 + 1) = 0, \tag{50.8}$$

namely,

$$\begin{vmatrix} a - r & h & f & l \\ h & b - r & g & m \\ f & g & c - r & n \\ l & m & n & d - r \end{vmatrix}. \tag{50.9}$$

When the coordinates are subjected to a transformation (30.11), equation (50.8) is transformed into

$$F'(x', y', z') - r(x'^2 + y'^2 + z'^2 + 1) = 0, \tag{50.10}$$

where $F'(x', y', z')$ is the transform of $F(x, y, z)$. The discriminant of (50.10) is

$$\begin{vmatrix} a' - r & h' & f' & l' \\ h' & b' - r & g' & m' \\ f' & g' & c' - r & n' \\ l' & m' & n' & d' - r \end{vmatrix}.$$

As remarked at the close of § 49, if the discriminant of the equation (50.8) is equal to zero, so also is the discriminant of equation (50.10). Equating these discriminants to zero, we have two equations of the fourth degree in r. Since these two equations have the same roots, the equations can differ at most by a constant factor. However, in each equation the coefficient of r^4 is $+1$; consequently corresponding coefficients are equal, and thus $\Delta = \Delta'$.

When a transformation (30.1) is applied to $F(x, y, z)$, and we denote the resulting expression by $F'(x', y', z')$, we have

$$\begin{aligned} F'(x', y', z') = {} & ax'^2 + by'^2 + cz'^2 + 2hx'y' + 2fx'z' + 2gy'z' \\ & + 2(ax_0 + hy_0 + fz_0 + l)x' + 2(hx_0 + by_0 + gz_0 + m)y' \\ & + 2(fx_0 + gy_0 + cz_0 + n)z' + F(x_0, y_0, z_0). \end{aligned}$$

The discriminant of $F'(x', y', z')$ is

$$\begin{vmatrix} a & h & f & ax_0 + hy_0 + fz_0 + l \\ h & b & g & hx_0 + by_0 + gz_0 + m \\ f & g & c & fx_0 + gy_0 + cz_0 + n \\ \begin{matrix} ax_0 + hy_0 \\ + fz_0 + l \end{matrix} & \begin{matrix} hx_0 + by_0 \\ + gz_0 + m \end{matrix} & \begin{matrix} fx_0 + gy_0 \\ + cz_0 + n \end{matrix} & F(x_0, y_0, z_0) \end{vmatrix}.$$

If we subtract from the last row the first multiplied by x_0, the second by y_0, and the third by z_0, we have, on considering $F(x_0, y_0, z_0)$ expressed in the form (49.2), that the above determinant is equal to

$$\begin{vmatrix} a & h & f & ax_0 + hy_0 + fz_0 + l \\ h & b & g & hx_0 + by_0 + gz_0 + m \\ f & g & c & fx_0 + gy_0 + cz_0 + n \\ l & m & n & lx_0 + my_0 + nz_0 + d \end{vmatrix},$$

which is readily seen to be equal to Δ. In view of these results and the fact that a general transformation is equivalent to a translation and a rotation, we have

[50.3] *The discriminant Δ of an equation of the second degree in three unknowns is an invariant under any change of rectangular coordinate axes.*

51. Classification of the Quadrics

We are now in position to analyze any equation of the second degree and determine the character of the surface defined by it and the position of the surface relative to the given coordinate axes.

In § 48 it was shown that given any such equation a coordinate system can be found in terms of which the equation is of the form

(51.1) $$a'x'^2 + b'y'^2 + c'z'^2 + 2\,l'x' + 2\,m'y' + 2\,n'z' + d' = 0.$$

In this coordinate system the characteristic equation (48.8) is

(51.2) $$\begin{vmatrix} a' - r & 0 & 0 \\ 0 & b' - r & 0 \\ 0 & 0 & c' - r \end{vmatrix} = 0,$$

of which the roots are a', b', c'. Thus when an equation of a quadric is in the general form (48.1), and one finds the roots of the characteristic equation (48.8), one has obtained the numbers which are the coefficients of the second-degree terms in an equation (51.1) into which the given equation may be transformed by a suitable transformation of coordinates.

In accordance with the theory of algebraic equations it follows from (48.9) that

(51.3) $$D = r_1 r_2 r_3,$$

where r_1, r_2, r_3 are the roots of the characteristic equation. We consider the two cases $D \neq 0$ and $D = 0$.

***Case 1.* $D \neq 0$.** By Theorem [**49.1**] the surface is an ellipsoid or hyperboloid when $\Delta \neq 0$, and a cone when $\Delta = 0$. If we effect a translation of axes to the center or vertex as new origin, it follows from equations (49.1) that $l = m = n = 0$ in the new system. Hence under such a translation equation (51.1) is transformed into an equation of the form

$$ax^2 + by^2 + cz^2 + d = 0.$$

For this equation we have from (49.4)

(51.4) $$\Delta = abcd = r_1 r_2 r_3 d.$$

Hence when $\Delta \neq 0$ an equation of the quadric is

(51.5) $$r_1 x^2 + r_2 y^2 + r_3 z^2 + \frac{\Delta}{r_1 r_2 r_3} = 0,$$

and when $\Delta = 0$ an equation is

(51.6) $$r_1 x^2 + r_2 y^2 + r_3 z^2 = 0,$$

where in each case r_1, r_2, r_3 are the roots of the characteristic equation. Hence we have

[51.1] *When for an equation of the second degree in x, y, and z $D \neq 0$ and $\Delta \neq 0$, the quadric is an ellipsoid, real or imaginary, when all the roots of the characteristic equation have the same sign, and a hyperboloid when the roots do not have the same sign; when $D \neq 0$ and $\Delta = 0$, the quadric is a cone or a point ellipsoid (imaginary cone) according as the roots have different signs or all have the same sign.*

If all the roots are equal, the quadric is a real or imaginary sphere when $\Delta \neq 0$, and a point sphere when $\Delta = 0$. If two and only two of the roots are equal and $D \neq 0$, the quadric is a surface of revolution.

Case 2. $D = 0$. One at least of the roots of the characteristic equation is equal to zero, as follows from (51.3). For the equation (51.1) this means that either a', b', or c' is equal to zero. Suppose that $c' = 0$ and that a' and b' are not zero. By a translation of the axes, as in § 47, equation (51.1) can be transformed into an equation of the form

$$ax^2 + by^2 + 2\,nz = 0, \tag{51.7}$$

or

$$ax^2 + by^2 + d = 0. \tag{51.8}$$

For equation (51.7)

$$\Delta = -\,abn^2 = -\,r_1 r_2 n^2, \tag{51.9}$$

and for (51.8) $\Delta = 0$. Hence when $\Delta \neq 0$ an equation of the quadric is

$$r_1 x^2 + r_2 y^2 + 2\sqrt{\frac{-\Delta}{r_1 r_2}}\,z = 0, \tag{51.10}$$

that is, the quadric is an elliptic or hyperbolic paraboloid according as the nonzero roots of the characteristic equation have the same or opposite signs. When $\Delta = 0$, we have (51.8) and an equation of the quadric is

$$r_1 x^2 + r_2 y^2 + d = 0. \tag{51.11}$$

By Theorem [**49.3**] the quadric is a cylinder or is degenerate according as equations (49.1) and (49.3) have not or have a common solution. It is an elliptic cylinder or two conjugate imaginary planes when the nonzero roots of the characteristic equation have the same sign; it is a hyperbolic cylinder or two real intersecting planes when the nonzero roots have opposite signs.

When two of the roots of (51.2) are equal to zero, we may in all generality take $b' = c' = 0$. As shown in § 47, the x'-, y'-, z'-coordinates can be chosen so that (51.1) is of the form

$$r_1 x^2 + 2\,my = 0, \tag{51.12}$$

or

$$r_1 x^2 + d = 0. \tag{51.13}$$

In both cases $\Delta = 0$, and by Theorem [**49.3**] the quadric is a parabolic cylinder or two parallel or coincident planes according as equations (49.1) and (49.3) have not or have a common solution.

Gathering these results together, we have

[**51.2**] *When for an equation of the second degree $D = 0$ and $\Delta \neq 0$, the quadric is a paraboloid, which is elliptic or hyperbolic according as the two nonzero roots of the characteristic equation have the same or different signs; when $D = 0$ and $\Delta = 0$ and the characteristic equation has two nonzero roots, the quadric is an elliptic or hyperbolic cylinder or consists of two intersecting planes, real or imaginary; when $D = 0$ and $\Delta = 0$ and there is only one nonzero root of the characteristic equation, the quadric is a parabolic cylinder or consists of parallel, or coincident, planes.*

The preceding results are set forth in the following table, in which a canonical form of an equation of a quadric is given in terms of the roots of the characteristic equation. When in a particular problem one has found the values of the roots, one is able to determine completely the form of the quadric, but in order to determine its position relative to the given coordinate axes it is necessary to find the principal planes by the method of § 48:

$$D \neq 0 \begin{cases} \Delta \neq 0 \quad r_1x^2 + r_2y^2 + r_3z^2 + \dfrac{\Delta}{r_1r_2r_3} = 0 \quad \ldots \text{ Ellipsoid or hyperboloid} \\ \Delta = 0 \quad r_1x^2 + r_2y^2 + r_3z^2 = 0 \quad \ldots\ldots \text{ Cone} \end{cases}$$

$$D = 0 \begin{cases} \Delta \neq 0 \quad r_1x^2 + r_2y^2 + 2\sqrt{\dfrac{-\Delta}{r_1r_2}}\, z = 0 \quad \ldots \text{ Paraboloid} \\ \Delta = 0 \begin{cases} r_1x^2 + r_2y^2 + d = 0 \begin{cases} d \neq 0 \ \ldots\ldots \text{ Elliptic or hyperbolic cylinder} \\ d = 0 \ \ldots\ldots \text{ Two planes} \end{cases} \\ d \neq 0, \text{ or } d = 0, \text{ as (49.1), (49.3) have not, or have, a common solution.} \\ r_1x^2 + 2\,my = 0 \quad \ldots\ldots \text{ Parabolic cylinder} \\ r_1x^2 + d = 0 \quad \ldots\ldots \text{ Two planes} \end{cases} \end{cases}$$

EXAMPLE 1. For the quadric

$$\text{(i)} \qquad x^2 + 2\,y^2 + 3\,z^2 - 4\,xy - 4\,yz + 2 = 0,$$

the characteristic equation is

$$r^3 - 6\,r^2 + 3\,r + 10 = 0,$$

of which the roots are 2, 5, -1. Since $\Delta = -20$, it follows from (51.5) that there is a coordinate system with respect to which an equation of the quadric is

$$\text{(ii)} \qquad 2\,x^2 + 5\,y^2 - z^2 + 2 = 0;$$

from this equation it follows that the quadric is a hyperboloid of two sheets. In order to find this new coordinate system, we observe that it follows from (49.1) that (0, 0, 0) is the center of the surface in the original coordinate system. Equations (48.7) for (i) are

$$\begin{aligned} (1 - r)u - 2\,v &= 0, \\ 2\,u + (r - 2)v + 2\,w &= 0, \\ 2\,v + (r - 3)w &= 0. \end{aligned}$$

Solutions of these equations for the roots 2, 5, -1 are 2, -1, -2; -1, 2, -2; 2, 2, 1 respectively. The lines through the origin with these respective direction numbers are the x-, y-, z-axes respectively of the coordinate system with respect to which (ii) is an equation of the hyperboloid. The three planes through the origin and whose normals have as direction numbers $2, -1, -2$; $-1, 2, -2$; $2, 2, 1$ in the original coordinate system are the principal planes of the hyperboloid.

EXAMPLE 2. For the quadric

(i) $$2x^2 + 2y^2 - 4z^2 - 5xy - 2xz - 2yz - 2x - 2y + z = 0$$

the characteristic equation is

$$r(r^2 - \tfrac{81}{4}) = 0,$$

of which the roots are $9/2, -9/2, 0$; and $D = 0$, $\Delta = 729/16$. From (51.10) we have that an equation of the surface is

(ii) $$3x^2 - 3y^2 + 2z = 0,$$

so that the surface is a hyperbolic paraboloid. For the roots $9/2$ and $-9/2$ respective solutions of (48.7) are $1, -1, 0$ and $1, 1, 4$. These are direction numbers in the original coordinate system of the x-axis and y-axis in the new system, and from (48.5) we have that in the original system equations of the new yz-plane and xz-plane are respectively

(iii) $$x - y = 0, \quad x + y + 4z = 0.$$

Since these two planes pass through the original origin, which is a point of the surface, and the new origin lies on the surface, as follows from (ii), the two origins coincide. Since the new xy-plane is perpendicular to the planes (iii) and passes through the original origin, an equation of the new xy-plane is

$$2x + 2y - z = 0.$$

EXAMPLE 3. For the quadric

(i) $$2x^2 - 2y^2 + 3z^2 + 3xy + 7xz - yz + 3x - 4y - z + d = 0$$

we have $D = 0$ and $\Delta = 0$ whatever be d. In this case the characteristic equation is

$$r(r^2 - 3r - \tfrac{75}{4}) = 0.$$

Since the two nonzero roots are different, the quadric is an elliptic or hyperbolic cylinder or two intersecting planes in accordance with Theorem [51.2]. Since $D = 0$, we have by

Theorem [41.5] that the portion of the above equation consisting of terms of the second degree is factorable. Factoring this expression, we consider the product

$$(2x - y + z + k)(x + 2y + 3z + m).$$

In order that this expression shall be the same as the left-hand member of equation (i), we must have

$$k + 2m = 3, \quad 2k - m = -4, \quad 3k + m = -1, \quad km = d.$$

The first three of these equations are satisfied by $k = -1$, $m = 2$. If then $d = km = -2$ in equation (i), the latter is an equation of a degenerate quadric, namely, two intersecting planes. If $d \neq -2$, equation (i) is an equation of a hyperbolic cylinder, since the nonzero roots of the characteristic equation differ in sign.

EXERCISES

Determine the form of each of the following quadrics and its relation to the coordinate axes:

1. $2x^2 + y^2 + 2z^2 + 2xy - 4x - 2y + 4z = 0.$

2. $x^2 + y^2 + 4z^2 - 2xy + 4xz - 4yz + 4x - 8z + 7 = 0.$

3. $x^2 + 2y^2 - z^2 + 4xy + 4xz - 2yz - 2x - 4z - 1 = 0.$

4. $x^2 - 2y^2 + z^2 + 2xy + 2xz + 2yz - 4x + 2y - 2z = 0.$

5. $x^2 + y^2 + z^2 + xy + xz + yz = 0.$

6. $2x^2 + 2y^2 + 3z^2 + 2xy - x + 6y + 6z + 9 = 0.$

7. $xy + 3xz + 14x + 2 = 0.$

8. $2x^2 - y^2 + z^2 - xy + 3xz - 2x + 2y - 2z = 0.$

9. $x^2 + 9y^2 + z^2 + 6xy - 2xz - 6yz - x - 3y + z = 0.$

10. Discuss the quadric of which the surface $\sqrt{x} - \sqrt{2y} + \sqrt{z} = 0$ is a part.

11. For what value of d is

$$x^2 - 2y^2 + 6z^2 + 12xz - 16x - 4y - 36z + d = 0$$

an equation of a cone?

12. For what value of a is $ax^2 + 6y^2 + 7z^2 + 4xz + 30 = 0$ an equation of a surface of revolution? Determine the form and position of the surface for this value of a.

13. Show that when for a general equation of the second degree $D = 0$, the portion of the equation consisting of terms of the second degree consists of two factors, distinct or not according as the characteristic equation has one zero root or two zero roots.

14. Show that for the quadric (48.1) the quantity $l^2 + m^2 + n^2$ is an invariant for any rotation of the coordinate axes.

15. Show that if three chords of a central quadric have the same mid-point, either all the chords lie in a plane or the common mid-point is the center of the quadric.

16. Find the most general equation of a quadric cone with vertex at the origin and having the x-, y-, and z-axes for elements.

17. Show that

$$ax^2 + 4\,y^2 + 9\,z^2 + 4\,xy + 6\,xz + 12\,yz + 2\,x - 2\,y + 2\,z + 8 = 0$$

is an equation of an elliptic paraboloid, a hyperbolic paraboloid, or a parabolic cylinder according as $a > 1$, $a < 1$, or $a = 1$.

18. Show that $a(x - y)^2 + b(x - z)^2 + c(y - z)^2 = d^2$ is an equation of an elliptic or hyperbolic cylinder according as $ab + bc + ca$ is positive or negative. Find equations of the line of points of symmetry in each case. Discuss the case when $ab + bc + ca = 0$.

19. Find the locus of the centers of the quadrics

$$x^2 + y^2 - z^2 + 2\,fxz + 2\,gyz + 2\,lx + 2\,my + 2\,nz = 0,$$

where l, m, and n are fixed numbers and f and g are parameters.

20. Three mutually perpendicular lines meet in a point P such that two of them intersect the axes of x and y respectively, and the third passes through a fixed point $(0, 0, c)$. Find the locus of P.

21. Consider equation (48.1) with $c \neq 0$ as a quadratic in z with x and y entering in the coefficients; solve for z and analyze the result after the manner of § 41.

Appendix to Chapter 1

In Chapter 1 we made use of definitions and theorems of Euclidean plane geometry to set up coordinate systems in which a point is represented by an ordered pair of numbers, and to prove that an equation of the first degree in two variables represents a (straight) line, in the sense that the coordinates of any point of a given line are a solution of a particular equation and every solution of this equation gives the coordinates of a point of this line. The advantage of this set-up is the ease with which it permits one to use algebra in the solution of geometric problems. However, the reader may appropriately raise the question: Do the methods of coordinate geometry enable one to solve any problem in Euclidean plane geometry? This question will be answered in the affirmative if we show that the processes which have been used imply a set of axioms from which all the theorems of plane geometry can be derived without making use (consciously or unconsciously) of any tacit assumption in deriving these theorems.

Before listing and testing such a set of axioms we shall sketch briefly the nature and development of the subject usually called Euclidean plane geometry. This subject has been the object of study from the eras of the intellectual glory of Egypt and of Greece. In the fourth century before the Christian Era, Euclid assembled in his *Elements* the results of the study of plane geometry by his predecessors and added to these results. The topic of Greek geometry was the description and investigation of different figures and their properties in the plane and in space. Some of the figures were encountered as shapes of material objects or generated by motion of bodies; others were produced by mechanical devices, and their properties were suggested by and derived from experience with physical phenomena.

It was the principal aim of Greek geometers to proceed with logical rigor, and this implied for them the necessity of deducing every geometric proposition from those previously established without undue reference to physical phenomena. However, every logical deduction must have a beginning somewhere. Accordingly, out of the accumulated knowledge and experience Euclid drew up a set of definitions and first propositions called axioms or postulates, and from these all subsequent propositions were derived by purely logical processes without further reference to physical intuition. (Euclid made a distinction between *axioms* and *postulates*, which, however, will be ignored in our discussion. We shall use the words interchangeably.) Euclid does not discuss the origin of the postulates or the philosophical means of deciding their validity. In his *Elements* he is concerned only with the possibility of deducing the known theorems of geometry from the postulates given, no matter what the origin of the latter might be.

For about two thousand years Euclid's *Elements* were accepted in most part as the final word in plane geometry. However, one point in Euclid was challenged from the earliest times; this was his parallel axiom: "If a straight line falling on two straight lines makes the interior angles on the same side of the line less than two right angles, the two straight lines if produced indefinitely meet on that side on which the angles are less than two right angles."

Most critics required that an axiom, since it is accepted without demonstration, should be sufficiently simple in its content to be self-evident, and the parallel axiom did not appear to be of that nature. For that reason many attempts were made to divest the proposition of its character as an axiom by proving it to be a logical consequence of the other axioms.

The most notable attempt was made by Saccheri (1733), who tried to prove the axiom by *reductio ad absurdum.* He tentatively replaced the parallel axiom by hypotheses opposed to it, with a view to deducing conclusions which would be obviously contradictory. The attempt was unsuccessful, but this type of argument prepared the way for geometries in which the parallel axiom would be replaced by an axiom opposed to it. The first concrete proposal was made independently about 1825 by the Russian mathematician Lobachevsky and the Hungarian Bolyai, namely, a system of geometry equally valid with the Euclidean results when Euclid's parallel postulate is replaced by the assumption that through a given point A not on a given line l there pass at least two lines which do not intersect l. This non-Euclidean geometry was studied in detail, and proof of its logical self-consistency is implied in the works of Cayley (1859). Consequently it is now known that any attempt to prove the parallel postulate as a consequence of the other Euclidean axioms must necessarily be futile.

These discoveries aroused anew widespread interest in plane geometry and its logical foundations, and new results led to a thorough overhauling of the axiomatic method. It was found that Euclid's set of first principles was in some respects incomplete and in other respects redundant. It was incomplete in so far as the results obtained by Euclid involved unannounced axioms because of the part played by intuition in obtaining these results. On the other hand, some of Euclid's definitions do not serve any useful mathematical purpose. For instance, Euclid states in the form of definitions: (1) a point is that which has no part; (2) a line is a breadthless length; (3) a straight line is a line which lies evenly with its points. These definitions can only be meant to announce to the reader that certain objects, as *point, line,* etc., will be studied henceforth. The description of them has no mathematical value and, as a matter of fact, Euclid never refers to these definitions in subsequent definitions or theorems. All significant information about these geometric objects is contained exclusively in the axioms, as, for instance, in the axiom that *any two points determine a straight line on which they lie.* Consequently it seems preferable to begin with the mere enumeration of certain mathematical objects by name and then to state axioms describing properties of these objects. The objects given at the outset are not defined explicitly, but the axioms state the relations which the objects have to one another. For instance, if we say that a point lies on a line, or that the line goes through the point, then we have an intuitive situation in mind. However, from the standpoint of mathematical rigor we must not refer to this intuition explicitly, but must realize that this fundamental relation is to be introduced and given by the axioms announcing it. Thus a modern set of axioms enumerates *undefined terms* and enunciates axioms containing these terms. (These undefined terms may designate either particular objects or classes of objects or relations.)

At the close of the last century the German mathematician Hilbert proposed a set of axioms, and since then has revised them. Many other mathematicians have likewise proposed sets of axioms. We give below a set of fifteen axioms for plane geometry as published by Hilbert in 1930 (*Grundlagen der Geometrie,* seventh edition), with some modifications for the sake of clarity.

The undefined terms in the axioms are six in number: *point, line, on* (a relation between a point and a line), *between* (a relation between a point and a pair of points), *congruent* (a relation between pairs of points), and *congruent* (a relation between angles).

As explanation but not as definition: The word "line" is here used to mean a straight line not terminated but extended indefinitely in both directions. A point B is said to be between the points A and C if it lies on the line AC and between A and C in the order of points on that line. A pair of points A, B is said to be congruent to a pair of points A', B' if the straight-line distance from A to B is equal to the straight-line distance from A' to B'. Two angles are said to be congruent if they are equal in measure, for example, in degrees, or if they are superposable one on the other. (As remarked, the statements in this paragraph are not *definitions,* but are *explanations* for the benefit of the reader who may be more familiar with these ideas under somewhat different names; from a logical point of view they not only can but should be omitted.)

Although congruence of angles is taken as an undefined term, the term "angle" is not itself undefined, but a definition is given for it below.

Following usual geometric terminology, in order to express that a point A is on (or lies on) a line l we may also say that *l passes through A* or that *l contains A*. We shall also use the phrase "the line AB" to designate the unique line which, according to Axiom 1, passes through A and B.

Appendix to Chapter 1

We now proceed to the statement of the fifteen axioms, interspersing them with definitions as necessary. The reader should draw a figure for each axiom as a means of clarifying its meaning.

Axiom 1. There is one and only one line passing through any two given (distinct) points.

Axiom 2. Every line contains at least two points, and given any line there is at least one point not on it.

Axiom 3. If a point B lies between the points A and C, then A, B, and C all lie on the same line, and B lies between C and A, and C does not lie between B and A, and A does not lie between B and C.

Axiom 4. Given any two (distinct) points A and C, there can always be found a point B which lies between A and C, and a point D such that C lies between A and D.

Axiom 5. If A, B, C are (distinct) points on the same line, one of the three points lies between the other two.

DEFINITION. The *segment* (or *closed interval*) AC consists of the points A and C and of all points which lie between A and C. A point B is said to be *on* the segment AC if it lies between A and C, or is A or C.

DEFINITION. Two lines, a line and a segment, or two segments, are said to *intersect* each other if there is a point which is on both of them.

DEFINITION. The *triangle* ABC consists of the three segments AB, BC, and CA (called the *sides* of the triangle), provided the points A, B, and C (called the *vertices* of the triangle) are not on the same line.

Axiom 6. A line which intersects one side of a triangle and does not pass through any of the vertices must also intersect one other side of the triangle.

Axiom 7. If A and B are (distinct) points and A' is a point on a line l, there exist two and only two (distinct) points B' and B'' on l such that the pair of points A', B' is congruent to the pair A, B and the pair of points A', B'' is congruent to the pair A, B; moreover A' lies between B' and B''.

Axiom 8. Two pairs of points congruent to the same pair of points are congruent to each other.

Axiom 9. If B lies between A and C, and B' lies between A' and C', and A, B is congruent to A', B', and B, C is congruent B', C', then A, C is congruent to A', C'.

DEFINITION. Two segments are *congruent* if their end points are congruent pairs of points.

DEFINITION. The *ray* AC consists of all points B which lie between A and C, the point C itself, and all points D such that C lies between A and D. (In consequence of preceding axioms it is readily proved that if C' is any point on the ray AC the rays AC' and AC are identical.) The ray AC is said to be *from* the point A.

DEFINITION. The *angle* BAC consists of the point A (the *vertex* of the angle) and the two rays AB and AC (the *sides* of the angle).

DEFINITION. If ABC is a triangle, the three angles BAC, ACB, CBA are called the *angles* of the triangle. Moreover the angle BAC is said to be *included* between the sides AB and AC of the triangle (and similarly for the other two angles of the triangle).

Axiom 10. If BAC is an angle whose sides do not lie in the same line, and B' and A' are (distinct) points, there exist two and only two (distinct) rays, $A'C'$ and $A'C''$, from A' such that the angle $B'A'C'$ is congruent to the angle BAC, and the angle $B'A'C''$ is congruent to the angle BAC; moreover if E' is any point on the ray $A'C'$ and E'' is any point on the ray $A'C''$, the segment $E'E''$ intersects the line $A'B'$.

Axiom 11. Every angle is congruent to itself.

Axiom 12. If two sides and the included angle of one triangle are congruent respectively to two sides and the included angle of another triangle, then the remaining angles of the first triangle are congruent each to the corresponding angle of the second triangle.

Axiom 13. Through a given point A not on a given line l there passes at most one line which does not intersect l.

Axiom 14. If A, B, C, D are (distinct) points, there exist on the ray AB a finite set of (distinct) points $A_1, A_2, \cdots, A_n$ such that (1) each of the pairs A, A_1; A_1, A_2; A_2, A_3; $\cdots$; A_{n-1}, A_n is congruent to the pair C, D and (2) B lies between A and A_n.

Axiom 15. The points of a line form a system of points such that no new points can be added to the space and assigned to the line without causing the line to violate one of the first eight axioms or Axiom 14.

The first five axioms state the simplest properties of a line and of the order possessed by points on a line. They assure in particular the existence of an endless number of points on a line, that a line is not terminated at any point, and that the order of points on a line is not like that of points on a closed curve such as a circle.

Axiom 6 also is concerned with order properties, but involves points not all on one line, and thus gives information about the plane as a whole in a way in which the previous axioms do not.

The congruence Axioms 7–12 are introduced so as to avoid in the proof of a proposition the use of superposition, that is, picking up a geometric figure and placing it upon another, as is done in the customary treatment of elementary geometry. Euclid himself used superposition, but even to Greek mathematicians and philosophers the use of this process in proving a theorem was open to question. Now mathematicians meet the question by means of congruence axioms. This explains, in particular, why Axioms 10 and 12 appear here as axioms and not as propositions to be proved.

Axiom 13 is the equivalent of Euclid's parallel axiom previously stated. It should be noted that Axioms 1–12 (because they lead to the essential properties of perpendicular lines) enable us to prove the existence of at least one line which passes through A and does not intersect l. Axiom 13 is required to assure us that there are not two such lines.

Axiom 14 is known as the Axiom of Archimedes. It corresponds to the process of using a measuring stick to find the distance from one point on a line to another, and insures that, starting at one point and laying off equal distances (the length of the stick) in succession towards the other point, the other point will ultimately be passed.

Axiom 15 is an axiom of completeness, assuring that there shall be on any line all the points necessary to constitute a continuum, which means that the points of any line may be brought into one-to-one correspondence with the set of all real numbers. In this sense Axiom 15 is equivalent to an axiom of continuity in the field of real numbers. For a discussion of this question the reader is referred to Part First of Fine's *College Algebra* and in particular to § 159, where the concept of a *Dedekind cut* is explained (but without calling it such). The question of continuity is fundamental in the calculus, and there the reader will find it fully discussed. Axiom 15 might be omitted from a set of axioms for elementary geometry since the usual elementary theorems follow without it, but it is necessary for the free use of real numbers in coordinate geometry, and especially for application of the calculus to geometry.

Granted these fifteen axioms, all further propositions of Euclidean plane geometry can be derived from them by a rigorous process of inference without further appeal to intuition. To carry this out in detail is, of course, a long story. The reader may consult in this connection O. Veblen's *The Foundations of Geometry,* Monographs on Topics of Modern Mathematics, pp. 3–51, Longmans, Green & Co., 1911, and H. G. Forder's *Foundations of Euclidean Geometry,* Cambridge University Press, 1927; in the latter many theorems of plane geometry are traced back individually to their axiomatic source, but there is much additional material.

Returning now to the question with which we introduced this Appendix, consider the following composite proposition (A), the six clauses of which (I–VI) correspond to the six undefined terms used in Axioms 1–15*:

Proposition A. It is possible to construct a coordinate system in the plane such that:

I. Every *point* has associated with it a unique pair of real numbers (its coordinates) and every pair of real numbers is associated with a unique point.

II. Every *line* has associated with it an equation $ax+by+c=0$ in which x and y are unknowns (variables), a, b, c are constant

* See in this connection O. Veblen's *The Modern Approach to Elementary Geometry,* The Rice Institute Pamphlet, Vol. 21, 1934, pp. 209–222.

coefficients, and a and b are not both 0. This equation is unique to within a possible constant multiplier ($\neq 0$), and every such equation is associated with a unique line.

III. A point is ***on*** a line, if and only if its coordinates satisfy an equation of the line.

IV. The point (x, y) lies ***between*** the points (x_1, y_1) and (x_2, y_2), if and only if there is a number t, greater than 0 and less than 1, such that the following equations both hold:

$$\begin{aligned} x &= (1 - t)x_1 + tx_2, \\ y &= (1 - t)y_1 + ty_2. \end{aligned} \tag{1}$$

V. The pair of points (x_1, y_1), (x_2, y_2) is ***congruent*** to the pair (x_3, y_3), (x_4, y_4), if and only if

$$(x_2 - x_1)^2 + (y_2 - y_1)^2 = (x_4 - x_3)^2 + (y_4 - y_3)^2. \tag{2}$$

VI. The angle $(x_2, y_2)(x_1, y_1)(x_3, y_3)$ is ***congruent*** to the angle $(x_2', y_2')(x_1', y_1')(x_3', y_3')$, if and only if

$$\begin{aligned} &\frac{(x_2 - x_1)(x_3 - x_1) + (y_2 - y_1)(y_3 - y_1)}{\sqrt{(x_2 - x_1)^2 + (y_2 - y_1)^2}\sqrt{(x_3 - x_1)^2 + (y_3 - y_1)^2}} \\ &= \frac{(x_2' - x_1')(x_3' - x_1') + (y_2' - y_1')(y_3' - y_1')}{\sqrt{(x_2' - x_1')^2 + (y_2' - y_1')^2}\sqrt{(x_3' - x_1')^2 + (y_3' - y_1')^2}}. \end{aligned} \tag{3}$$

Proposition A, I necessitates Axiom 15 since, as previously remarked, this axiom guarantees the existence upon any line of points in one-to-one correspondence with all real numbers. As a matter of fact, points corresponding to a suitable subset of real numbers satisfy Axioms 1–14, and such a subset would suffice to give algebraic expression to the fourteen axioms. However, we desire to deal with all real numbers and thus have included Axiom 15.

The truth of Proposition A follows as a consequence of familiar theorems of elementary Euclidean geometry (see Theorems [**5.2**], [**5.3**], [**3.1**], [**3.4**] and equations (5.8)).

We shall now show conversely that Axioms 1–14 follow as consequences of Proposition A entirely by the ***methods of algebra*** and without use of other propositions of geometry. We call the left-hand member of equation (2) the square of the length of the line segment $(x_1, y_1)(x_2, y_2)$, and the left-hand member of equation (3) the cosine of the angle $(x_2, y_2)(x_1, y_1)(x_3, y_3)$. These names are applied to algebraic expressions involving x_1, y_1, and so on, not to geometric objects.

Axioms 1 and 2 follow from the algebraic definitions of point and line in A and the results of § 1.

For any real value of $t > 1$ equations (1) determine a point. On solving these equations for x_2 and y_2, we obtain

$$x_2 = \frac{1}{t}x + \left(1 - \frac{1}{t}\right)x_1, \quad y_2 = \frac{1}{t}y + \left(1 - \frac{1}{t}\right)y_1,$$

from which it is seen that (x_2, y_2) lies between (x, y) and (x_1, y_1), in accordance with the algebraic definition A, IV of betweenness. Similarly, it can be shown that if $t < 0$, (x_1, y_1) lies between (x, y) and (x_2, y_2). Thus Axioms 3, 4, and 5 follow from the algebraic definition A, IV.

Theorem [6.8] gives an algebraic definition of parallelism satisfying Axiom 13. From this definition and the algebraic definition of direction numbers of a line in terms of the coordinates of any two points of the line, there follows the result that direction numbers of parallel lines are proportional.

In the consideration of Axiom 6 we take on the side P_1P_2 of the triangle with vertices $P_1(x_1, y_1)$, $P_2(x_2, y_2)$, $P_3(x_3, y_3)$ a point $P(x, y)$ other than P_1 and P_2. Its coordinates are given by (1) for $0 < t < 1$. If the line of the axiom is parallel to P_2P_3, it meets the side P_1P_3 in the point P' of coordinates $(1 - t)x_1 + tx_3$, $(1 - t)y_1 + ty_3$, since direction numbers of the line segment PP' are $t(x_3 - x_2)$, $t(y_3 - y_2)$, which are direction numbers of P_2P_3. Similar results hold for a line parallel to P_1P_3. Suppose then that the line through P is not parallel to P_1P_3 or P_2P_3, and denote by P' and P'' the points in which it meets the lines P_1P_3 and P_2P_3 respectively. The coordinates of P' are of the form

$$(1 - r)x_1 + rx_3, \quad (1 - r)y_1 + ry_3.$$

Since the line does not pass through P_1 or P_3, r is not equal to 0 or 1. If r lies between 0 and 1, the conditions of the axiom are met. Suppose then that $r < 0$ or > 1, in which case P' lies outside the segment P_1P_3, and write the coordinates of P'' thus:

$$(1 - s)x_2 + sx_3, \quad (1 - s)y_2 + sy_3.$$

Since the points P, P', P'' lie on a line, there exists a number u such that (see page 25)

$$(1 - s)x_2 + sx_3 = (1 - u)[(1 - t)x_1 + tx_2] + u[(1 - r)x_1 + rx_3],$$

and similarly for the y's. Since P_1, P_2, P_3 are not collinear, there can be no linear homogeneous relation of their coordinates, and consequently we must have

$$(1-u)(1-t)+u(1-r)=0, \quad (1-u)t-(1-s)=0, \quad ur-s=0.$$

The first of these equations is satisfied if the other two are. Eliminating u from the second and third of these equations, we obtain

$$s = \frac{1 - t}{1 - \dfrac{t}{r}}.$$

Since $0 < t < 1$, and $r < 0$ or > 1, by hypothesis, we have that $0 < s < 1$; that is, P'' is on the segment P_2P_3; and the axiom follows.

Axiom 8 is satisfied by the algebraic definition A, V of congruence of pairs of points. If (x_1, y_1) and (x_2, y_2) are the points A and B of Axiom 7, (x_3, y_3) the point A' on the line l, and $a(x - x_3) + b(y - y_3) = 0$ an equation of the line l, the determination of the points B' and B'' is the algebraic problem of finding the common solutions of the equations

$$(x_2 - x_1)^2 + (y_2 - y_1)^2 = (x - x_3)^2 + (y - y_3)^2$$
$$a(x - x_3) + b(y - y_3) = 0.$$

It is evident that these equations have two and only two solutions; and one can show that (x_3, y_3) is the mid-point between the points B' and B'' (see Theorem [**4.1**]).

In the consideration of Axiom 9, we denote by (x_1, y_1), (x_2, y_2), (x_1', y_1'), (x_2', y_2') the coordinates of A, C, A', C' respectively. Then B and B' are points (x, y) and (x', y'), where

$$x = (1 - t)x_1 + tx_2, \qquad y = (1 - t)y_1 + ty_2,$$
$$x' = (1 - t')x_1' + t'x_2', \quad y' = (1 - t')y_1' + t'y_2',$$

for suitable values of t and t' such that $0 < t < 1$, $0 < t' < 1$.

When now in accordance with A,V we express the congruence of AB and $A'B'$, and of BC and $B'C'$, we find that $t' = t$, and then that AC and $A'C'$ are congruent; and the axiom follows.

If for Axiom 10 the angle BAC is $(x_2, y_2)(x_1, y_1)(x_3, y_3)$ and the points A', B' are (x_1', y_1'), (x_2', y_2'), the left-hand member of equation (3) is a fixed number k which can be shown to be such that $|k| < 1$, and we have to determine solutions x_3', y_3' of this equation. If we put

$$d_2 = \sqrt{(x_2'-x_1')^2+(y_2'-y_1')^2}, \quad u_2 = \frac{x_2'-x_1'}{d_2}, \quad v_2 = \frac{y_2'-y_1'}{d_2},$$

then

$$(4) \qquad x_2' = x_1' + d_2u_2, \qquad y_2' = y_1' + d_2v_2;$$

and for any positive value of d_2 these equations give the coordinates of a point on the ray $A'B'$, and for negative values of d_2 on the ray through A' in the opposite direction. If, in like manner, we put

$$d = \sqrt{(x'-x_1')^2+(y'-y_1')^2}, \quad u = \frac{x'-x_1'}{d}, \quad v = \frac{y'-y_1'}{d},$$

and require x', y' to satisfy equation (3), we obtain

$$k = uu_2 + vv_2.$$

If this equation is written in the form

$$(5) \qquad vv_2 = k - uu_2$$

and then squared, the resulting equation is reducible to

$$(6) \qquad (u - u_2k)^2 = \sqrt{(1-k^2)v_2^2},$$

since

$$u_2^2 + v_2^2 = 1, \qquad u^2 + v^2 = 1.$$

If we put $l = \sqrt{1-k^2}$, we have from (5) and (6)

$$u = u_2k \pm v_2l, \qquad v = v_2k \mp u_2l.$$

Consequently there are two and only two solutions of the problem. The coordinates of points E' and E'' on the respective rays from A' forming with $A'B'$ an angle congruent to the given angle are given by

$$x_3' = x_1' + d_3(u_2k + v_2l), \qquad y_3' = y_1' + d_3(v_2k - u_2l),$$
$$x_4' = x_1' + d_4(u_2k - v_2l), \qquad y_4' = y_1' + d_4(v_2k + u_2l),$$

for positive values of d_3 and d_4. Since

$$\left(1 - \frac{d_3}{d_3 + d_4}\right)x_3' + \frac{d_3}{d_3 + d_4}\,x_4' = x_1' + \frac{d_3 d_4}{d_3 + d_4}\,ku_2,$$

$$\left(1 - \frac{d_3}{d_3 + d_4}\right)y_3' + \frac{d_3}{d_3 + d_4}\,y_4' = y_1' + \frac{d_3 d_4}{d_3 + d_4}\,kv_2,$$

it follows from (4) by A, IV that the segment $E'E''$ has a point in common with the ray $A'B'$ when k is positive, and with the opposite of the ray when k is negative. Thus Axiom 10 follows from A, V and A, IV as Axiom 11 does directly from A, V.

We denote by l_2^2 the left-hand member of equation (2), which we have called the square of the length of the line segment $(x_1, y_1)(x_2, y_2)$, and similarly we denote by l_1^2 and l^2 the squares of the lengths of the segments $(x_1, y_1)(x_3, y_3)$ and $(x_2, y_2)(x_3, y_3)$ respectively; also we denote by $\cos A$ the left-hand member of equation (3), which we have called the cosine of the angle $(x_2, y_2)(x_1, y_1)(x_3, y_3)$. When these expressions are substituted in the equation

$$l^2 = l_1^2 + l_2^2 - 2\,l_1 l_2 \cos A, \tag{7}$$

it is found that this equation is an identity, that is, the Law of Cosines is thus an algebraic identity. If in Axiom 12 the vertices of the two triangles are (x_1, y_1), (x_2, y_2), (x_3, y_3) and (x_1', y_1'), (x_2', y_2'), (x_3', y_3'), the expressions (2) for the lengths $(x_1, y_1)(x_2, y_2)$ and $(x_1', y_1')(x_2', y_2')$ are equal, and likewise for $(x_1, y_1)(x_3, y_3)$ and $(x_1', y_1')(x_3', y_3')$. The expressions (3) for $\cos A$ and $\cos A'$ of the angles at (x_1, y_1) and (x_1', y_1') are equal. Then from equation (7) we have the equality of the expressions for the lengths of the third sides. Having the expressions for the lengths of corresponding sides of the triangles equal, we obtain from equations of the form (7) the equality of the expressions for the cosines of corresponding angles and by A, VI the congruence of these angles.

If in Axiom 14 the coordinates of A and B are x_1, y_1 and x_2, y_2 respectively, and the coordinates of A_1 are given by (1), then t is equal to l_1/l_2, where l_1 and l_2 are the lengths of the segments CD and AB respectively. If then we replace t in (1) by $2\,t$, $3\,t$, and so on, equations (1) give the coordinates of the points A_2, A_3, and so on, of the axiom. Consequently this axiom is satisfied by taking a sufficiently large integer n as multiplier of t in equations (1).

Since it has now been shown that Axioms 1–15 follow as consequences of Proposition A *entirely by the methods of algebra* and without use of other propositions of geometry, the situation is as follows:

Clauses I–VI of Proposition A provide algebraic correspondents for each of the six undefined geometric terms. Given any geometric term, write out its definition from the six undefined terms, and in the definition replace each undefined term by its algebraic correspondent; the result will be the algebraic correspondent of the given geometric term. Hence every geometric term has its algebraic correspondent or representative. This enables one to translate every geometric theorem into a corresponding theorem of algebra. And in view of the algebraic proofs provided for the theorems of algebra corresponding to Axioms 1–15, the geometric proof of a theorem can be translated into an algebraic proof of the corresponding theorem of algebra (although, as indeed often happens, a shorter algebraic proof may be found in another way).

Thus the question which we raised concerning the adequacy of the methods of coordinate geometry in the study of any question of Euclidean plane geometry is answered in the affirmative.

Index

Index

Index

Index

A CATALOG OF SELECTED

DOVER BOOKS

IN SCIENCE AND MATHEMATICS

Math–Geometry and Topology

ELEMENTARY CONCEPTS OF TOPOLOGY, Paul Alexandroff. Elegant, intuitive approach to topology from set-theoretic topology to Betti groups; how concepts of topology are useful in math and physics. 25 figures. 57pp. 5⅜ x 8½. 60747-X

COMBINATORIAL TOPOLOGY, P. S. Alexandrov. Clearly written, well-organized, three-part text begins by dealing with certain classic problems without using the formal techniques of homology theory and advances to the central concept, the Betti groups. Numerous detailed examples. 654pp. 5⅜ x 8½. 40179-0

EXPERIMENTS IN TOPOLOGY, Stephen Barr. Classic, lively explanation of one of the byways of mathematics. Klein bottles, Moebius strips, projective planes, map coloring, problem of the Koenigsberg bridges, much more, described with clarity and wit. 43 figures. 210pp. 5⅜ x 8½. 25933-1

CONFORMAL MAPPING ON RIEMANN SURFACES, Harvey Cohn. Lucid, insightful book presents ideal coverage of subject. 334 exercises make book perfect for self-study. 55 figures. 352pp. 5⅝ x 8¼. 64025-6

THE GEOMETRY OF RENÉ DESCARTES, René Descartes. The great work founded analytical geometry. Original French text, Descartes's own diagrams, together with definitive Smith-Latham translation. 244pp. 5⅜ x 8½. 60068-8

PRACTICAL CONIC SECTIONS: The Geometric Properties of Ellipses, Parabolas and Hyperbolas, J. W. Downs. This text shows how to create ellipses, parabolas, and hyperbolas. It also presents historical background on their ancient origins and describes the reflective properties and roles of curves in design applications. 1993 ed. 98 figures. xii+100pp. 6½ x 9¼. 42876-1

THE THIRTEEN BOOKS OF EUCLID'S ELEMENTS, translated with introduction and commentary by Thomas L. Heath. Definitive edition. Textual and linguistic notes, mathematical analysis. 2,500 years of critical commentary. Unabridged. 1,414pp. 5⅜ x 8½.. Three-vol. set. Vol. I: 60088-2 Vol. II: 60089-0 Vol. III: 60090-4

GEOMETRY OF COMPLEX NUMBERS, Hans Schwerdtfeger. Illuminating, widely praised book on analytic geometry of circles, the Moebius transformation, and two-dimensional non-Euclidean geometries. 200pp. 5⅝ x 8¼. 63830-8

DIFFERENTIAL GEOMETRY, Heinrich W. Guggenheimer. Local differential geometry as an application of advanced calculus and linear algebra. Curvature, transformation groups, surfaces, more. Exercises. 62 figures. 378pp. 5⅜ x 8½. 63433-7

CURVATURE AND HOMOLOGY: Enlarged Edition, Samuel I. Goldberg. Revised edition examines topology of differentiable manifolds; curvature, homology of Riemannian manifolds; compact Lie groups; complex manifolds; curvature, homology of Kaehler manifolds. New Preface. Four new appendixes. 416pp. 5⅜ x 8½. 40207-X

Mathematics

FUNCTIONAL ANALYSIS (Second Corrected Edition), George Bachman and Lawrence Narici. Excellent treatment of subject geared toward students with background in linear algebra, advanced calculus, physics, and engineering. Text covers introduction to inner-product spaces, normed, metric spaces, and topological spaces; complete orthonormal sets, the Hahn-Banach Theorem and its consequences, and many other related subjects. 1966 ed. 544pp. 6⅛ x 9¼. 40251-7

ASYMPTOTIC EXPANSIONS OF INTEGRALS, Norman Bleistein & Richard A. Handelsman. Best introduction to important field with applications in a variety of scientific disciplines. New preface. Problems. Diagrams. Tables. Bibliography. Index. 448pp. 5⅜ x 8½. 65082-0

VECTOR AND TENSOR ANALYSIS WITH APPLICATIONS, A. I. Borisenko and I. E. Tarapov. Concise introduction. Worked-out problems, solutions, exercises. 257pp. 5⅝ x 8¼. 63833-2

THE ABSOLUTE DIFFERENTIAL CALCULUS (CALCULUS OF TENSORS), Tullio Levi-Civita. Great 20th-century mathematician's classic work on material necessary for mathematical grasp of theory of relativity. 452pp. 5⅝ x 8¼. 63401-9

AN INTRODUCTION TO ORDINARY DIFFERENTIAL EQUATIONS, Earl A. Coddington. A thorough and systematic first course in elementary differential equations for undergraduates in mathematics and science, with many exercises and problems (with answers). Index. 304pp. 5⅜ x 8½. 65942-9

FOURIER SERIES AND ORTHOGONAL FUNCTIONS, Harry F. Davis. An incisive text combining theory and practical example to introduce Fourier series, orthogonal functions and applications of the Fourier method to boundary-value problems. 570 exercises. Answers and notes. 416pp. 5⅜ x 8½. 65973-9

COMPUTABILITY AND UNSOLVABILITY, Martin Davis. Classic graduate-level introduction to theory of computability, usually referred to as theory of recurrent functions. New preface and appendix. 288pp. 5⅜ x 8½. 61471-9

ASYMPTOTIC METHODS IN ANALYSIS, N. G. de Bruijn. An inexpensive, comprehensive guide to asymptotic methods–the pioneering work that teaches by explaining worked examples in detail. Index. 224pp. 5⅜ x 8½ 64221-6

APPLIED COMPLEX VARIABLES, John W. Dettman. Step-by-step coverage of fundamentals of analytic function theory–plus lucid exposition of five important applications: Potential Theory; Ordinary Differential Equations; Fourier Transforms; Laplace Transforms; Asymptotic Expansions. 66 figures. Exercises at chapter ends. 512pp. 5⅜ x 8½. 64670-X

INTRODUCTION TO LINEAR ALGEBRA AND DIFFERENTIAL EQUATIONS, John W. Dettman. Excellent text covers complex numbers, determinants, orthonormal bases, Laplace transforms, much more. Exercises with solutions. Undergraduate level. 416pp. 5⅜ x 8½. 65191-6

CALCULUS OF VARIATIONS WITH APPLICATIONS, George M. Ewing. Applications-oriented introduction to variational theory develops insight and promotes understanding of specialized books, research papers. Suitable for advanced undergraduate/graduate students as primary, supplementary text. 352pp. 5⅜ x 8½. 64856-7

COMPLEX VARIABLES, Francis J. Flanigan. Unusual approach, delaying complex algebra till harmonic functions have been analyzed from real variable viewpoint. Includes problems with answers. 364pp. 5⅜ x 8½. 61388-7

AN INTRODUCTION TO THE CALCULUS OF VARIATIONS, Charles Fox. Graduate-level text covers variations of an integral, isoperimetrical problems, least action, special relativity, approximations, more. References. 279pp. 5⅜ x 8½. 65499-0

COUNTEREXAMPLES IN ANALYSIS, Bernard R. Gelbaum and John M. H. Olmsted. These counterexamples deal mostly with the part of analysis known as "real variables." The first half covers the real number system, and the second half encompasses higher dimensions. 1962 edition. xxiv+198pp. 5⅜ x 8½. 42875-3

CATASTROPHE THEORY FOR SCIENTISTS AND ENGINEERS, Robert Gilmore. Advanced-level treatment describes mathematics of theory grounded in the work of Poincaré, R. Thom, other mathematicians. Also important applications to problems in mathematics, physics, chemistry, and engineering. 1981 edition. References. 28 tables. 397 black-and-white illustrations. xvii+666pp. 6⅛ x 9¼. 67539-4

INTRODUCTION TO DIFFERENCE EQUATIONS, Samuel Goldberg. Exceptionally clear exposition of important discipline with applications to sociology, psychology, economics. Many illustrative examples; over 250 problems. 260pp. 5⅜ x 8½. 65084-7

NUMERICAL METHODS FOR SCIENTISTS AND ENGINEERS, Richard Hamming. Classic text stresses frequency approach in coverage of algorithms, polynomial approximation, Fourier approximation, exponential approximation, other topics. Revised and enlarged 2nd edition. 721pp. 5⅜ x 8½. 65241-6

INTRODUCTION TO NUMERICAL ANALYSIS (2nd Edition), F. B. Hildebrand. Classic, fundamental treatment covers computation, approximation, interpolation, numerical differentiation and integration, other topics. 150 new problems. 669pp. 5⅜ x 8½. 65363-3

THREE PEARLS OF NUMBER THEORY, A. Y. Khinchin. Three compelling puzzles require proof of a basic law governing the world of numbers. Challenges concern van der Waerden's theorem, the Landau-Schnirelmann hypothesis and Mann's theorem, and a solution to Waring's problem. Solutions included. 64pp. 5⅜ x 8½. 40026-3

THE PHILOSOPHY OF MATHEMATICS: An Introductory Essay, Stephan Körner. Surveys the views of Plato, Aristotle, Leibniz & Kant concerning propositions and theories of applied and pure mathematics. Introduction. Two appendices. Index. 198pp. 5⅜ x 8½. 25048-2